Direct Nerve Stimulation for Induction of Sensation and Treatment of Phantom Limb Pain

RIVER PUBLISHERS SERIES IN BIOMEDICAL ENGINEERING

Indexing: All books published in this series are submitted to the Web of Science Book Citation Index (BkCI), to SCOPUS, to CrossRef and to Google Scholar for evaluation and indexing.

The "River Publishers Series in Biomedical Engineering" is a series of comprehensive academic and professional books which focus on the engineering and mathematics in medicine and biology. The series presents innovative experimental science and technological development in the biomedical field as well as clinical application of new developments.

Books published in the series include research monographs, edited volumes, handbooks and textbooks. The books provide professionals, researchers, educators, and advanced students in the field with an invaluable insight into the latest research and developments.

Topics covered in the series include, but are by no means restricted to the following:

- Biomedical engineering
- Biomedical physics and applied biophysics
- Bio-informatics
- Bio-metrics
- Bio-signals
- Medical Imaging

For a list of other books in this series, visit www.riverpublishers.com

Direct Nerve Stimulation for Induction of Sensation and Treatment of Phantom Limb Pain

Editor

Winnie Jensen

Aalborg University, Denmark

Published 2019 by River Publishers
River Publishers
Alsbjergvej 10, 9260 Gistrup, Denmark
www.riverpublishers.com

Distributed exclusively by Routledge
4 Park Square, Milton Park, Abingdon, Oxon OX14 4RN
605 Third Avenue, New York, NY 10158

First published in paperback 2024

Direct Nerve Stimulation for Induction of Sensation and Treatment of Phantom Limb Pain / by Winnie Jensen.

Routledge is an imprint of the Taylor & Francis Group, an informa business

ISBN: 978-87-7022-076-7 (hbk)
ISBN: 978-87-7004-362-5 (pbk)
ISBN: 978-1-003-33797-3 (ebk)

DOI: 10.1201/9781003337973

Contents

Preface

Phantom limb pain (PLP) is a frequent consequence of amputation, and it is notoriously difficult to treat. Despite isolated reports of success, no medical/non-medical treatments have been beneficial on more than a temporary basis. While the majority of the treatments currently offered seek to actively suppress the pain, we emabarked on a journey back in 2008 to challenge the status-quo of PLP treatment by instead supplying meaningful sensations that will restore the neuroplastic changes in the cortex and thereby control and alleviate pain. We designed, implemented and tested a novel 'human-machine interface' that included a 'first-in-human' clinical trial of the system. In this book we report on the first steps and results from this journey to demonstrate and provide a proof of concept of our ideas.

Acknowledgements

The authors would like to thanks the European Commission for the support for this work, which was primarily funded through the CP-FP-INFSO 224012 (TIME project). The views and ideas expressed in this book are those of the authors and do not necessarily represent those of the European Commission. This work was partially supported by national funds through BEVICA fonden (DK), FEDER and CIBERNED funds Fondo de Investigación Sanitaria of Spain, the FET 611687 NEBIAS Project, the IUPUI BME Department Research Assistantship Fund and The National Competence Center in Research (NCCR) in Robotics funded by the Swiss National Science Foundation, and the Bertarelli Foundation.

June 2019, Winnie Jensen

List of Contributors

Aritra Kundu, *Department of Health Science and Technology, Aalborg University, Denmark*

Bo Geng, *SMI, Department of Health Science and Technology, Aalborg University, Denmark*

Caleb C. Comoglio, *Department of Biomedical Engineering, Indiana University – Purdue University Indianapolis*

David Andreu, *LIRMM, University of Montpellier, INRIA, CNRS, Montpellier, France*

David Guiraud, *LIRMM, University of Montpellier, INRIA, CNRS, Montpellier, France*

Elena Redondo-Castro, *Institute of Neurosciences, Department of Cell Biology, Physiology and Immunology, Universitat Autònoma de Barcelona, and Centro de Investigación Biomédica en Red Sobre Enfermedades Neurodegenerativas (CIBERNED), Bellaterra, Spain*

Giacomo Valle, *Bertarelli Foundation Chair in Translational Neuroengineering, Centre for Neuroprosthetics and Institute of Bioengineering, School of Engineering, École Polytechnique Fédérale de Lausanne (EPFL), Lausanne, Switzerland;*
The BioRobotics Institute, Scuola Superiore Sant'Anna, Pisa, Italy

Guillaume Souquet, *Axonic, MXM, France*

Guiseppe Granata, *Fondazione Policlinico Universitario A. Gemelli – IRCCS, Italy;*
Università Cattolica del Sacro Cuore, Italy;
E-mail: granata.gius@gmail.com

Guy Cathebras, *LIRMM, University of Montpellier, INRIA, CNRS, Montpellier, France*

Jean-Louis Divoux, *Axonic, MXM, France*

Jordi Badia, *Institute of Neurosciences and Department of Cell Biology, Physiology and Immunology, Universitat Autònoma de Barcelona, Bellaterra, Spain;*
Centro de Investigación Biomédica en Red sobre Enfermedades Neurodegenerativas (CIBERNED), Spain

Ken Yoshida, *Department of Biomedical Engineering, Indiana University – Purdue University Indianapolis;*
E-mail: yoshidak@iupui.edu

Kristian R. Harreby, *Department of Health Science and Technology, Aalborg University, Denmark*

Kristine Mosier, *Department of Radiology and Imaging Sciences, Indiana University School of Medicine*

Loic Wauters, *Axonic, MXM, France*

P. M. Rossini, *Fondazione Policlinico Universitario A. Gemelli – IRCCS, Italy;*
Università Cattolica del Sacro Cuore, Italy

Pawel Maciejasz, *LIRMM, University of Montpellier, INRIA, CNRS, Montpellier, France*

Robin Passama, *LIRMM, University of Montpellier, INRIA, CNRS, Montpellier, France*

Silvestro Micera, *Bertarelli Foundation Chair in Translational Neuroengineering, Centre for Neuroprosthetics and Institute of Bioengineering, School of Engineering, École Polytechnique Fédérale de Lausanne (EPFL), Lausanne, Switzerland;*
The BioRobotics Institute, Scuola Superiore Sant'Anna, Pisa, Italy;
E-mail: silvestro.micera@epfl.ch

Thomas Stieglitz, *Laboratory for Biomedical Microsystems, Department of Microsystems Engineering-IMTEK, Albert-Ludwig-University of Freiburg, Freiburg, Germany;*
BrainLinks-BrainTools, Albert-Ludwig-University of Freiburg, Freiburg, Germany;
Bernstein Center Freiburg, Albert-Ludwig-University of Freiburg, Freiburg, Germany;
E-mail: stieglitz@imtek.uni-freiburg.de

Tim Boretius, *Neuroloop GmbH, Freiburg, Germany;*
Laboratory for Biomedical Microsystems, Department of Microsystems Engineering-IMTEK, Albert Ludwig University of Freiburg, Freiburg, Germany

Víctor M. López-Álvarez, *Institute of Neurosciences, Department of Cell Biology, Physiology and Immunology, Universitat Autònoma de Barcelona, and Centro de Investigación Biomédica en Red Sobre Enfermedades Neurodegenerativas (CIBERNED), Bellaterra, Spain*

Winnie Jensen, *Department of Health Science and Technology, Aalborg University, Denmark;*
E-mail: wj@hst.aau.dk

Xavier Navarro, *Institute of Neurosciences, Department of Cell Biology, Physiology and Immunology, Universitat Autònoma de Barcelona, Bellaterra, Spain;*
Centro de Investigación Biomédica en Red Sobre Enfermedades Neurodegenerativas (CIBERNED), Spain;
E-mail: xavier.navarro@uab.cat

List of Figures

List of Tables

List of Abbreviations

AG	assembly groups
AP	action potential
API	application programming interface
AS	active sites
ASICS	acid sensing channels
ASIP	application-specific instruction-set processor
BOLD	Blood oxygen level dependent
BPA	brachial plexus avulsion
BPI	brief pain inventory
BPI-IS	Brief Pain Inventory – Interference Scale
BPI-SF	brief pain inventory – short form
CES-D	Center for Epidemiological Studies – depression questionnaire
CGRP	calcitonin gene-related peptide
CMAP	compound muscle action potentials
CNAP	compount nerve action potentials
CNS	Central nervous system
CPG	chronic pain grade
CRPS	complex regional pain syndrome
cVLM	caudal ventrolateral medulla
cw	constant weigthing
d	device
DACC	digital to analog current converter
DBS	deep brain stimulation
Deg/ENaC	degenerin family
DN4	the neuropathic pain four questions
DNIC	diffuse noxious inhibitory control
DRt	dorsal reticular nucleus
EDM	electrical discharge machining
EEG	electroencephalogram
EES	epidural eletrical stimulation
EIS	electrochemical impedance spectroscopy

EMG	electromyogram
ES	electrical stimulation
FBR	foreign body response
FEM	finite element method
FEP	flourinated ethylene propylene
FES	functional electrical stimulation
FFC	flexible flat cable
FIR	finite impulse response
fMRI	functional magnetic resonance imaging
FPGA	programmable electronic device
FRAP	non-peptidergic ones possess fluoride-resistant acid phosphatase
FTIR	fourier transform infrared spectroscopy
GALS	Globally Asynchronous Locally Synchronous
GDNF	glial cell line-derived neurotrophic factor
GM	gastrocnemious medialis
GMI	Graded motor imagery
GND	ground
GQPAA	Groningen questionnaire problems after arm amputation
GUI	graphical user interface
H&E	Hematoxylin and Eosin
HCNS	heterotopic noxious conditioning stimulation
HCP	health care provider
HMI	Human machine interface
HRF	hemodynamic response function
IASP	International Association for the Study of Pain
ICA	indpendent component analysis
IMMPACT	the initiative on methods, measurement and pain assessment in clinical trails
IPA	isopropyl alcohol
ISI	interactive subject interface
LANSS	Leeds assessment of neuropathic symptoms and signs
LEF	laboratory for electrode manufacturing
LEP	laser evoked potential
LIFE	Longitudinal Intrafascicular electrode
LTD	long-term depression
LTP	long-term potentiation
M1	primary motor cortex
MAC	medium access control

MAV	mean absolute value
MFI	microflex interconnector
MP	micro program
MPI	West Haven-Yale multidimensional pain inventory
MPQ	McGill pain question
NGF	nerve growth factor
NMDA	N-methyl D-aspartate
NOS	NO synthase
NP	neuropathic pain
NPC	nano plastic circular
NPQ	neuropathic pain questionnaire
NPS	Neuropathic pain scale
NPSI	neuropathic pain symptom inventory
NRS	numeric rating scale
PAG	periaqueductal gray matter
PAP	Post amputation pain
PBS	phosphate buffered saline
PECVD	plasma enhanced chemical vapour deposition
PEI	polyesterimide
PEQ	prosthesis evaluation questionnaire
PKCγ	protein kinase Cgamma
PL	plantar interosseus
PLP	Phantom limb pain
PLS	Phantom limb sensation
PN	petri nets
PNS	peripheral nervous system
POMS-SF	profile of mood states – short form
PPI	present pain intensity
Pt	platinum
PVCN	posteroventral cochlear nucleus
QMS	quality management system
RCT	randomized controlled trials
RIE	reactive ion etching
RL	recruitment level
RLP	Residual limb pain
RMN	raphe magnocellular nucleus
RMS	root mean square
ROI	region of interest
RS	referred sensation

S1	Primary somatosensory cortex
S2	Secondary somatosensory cortex
SEC	stimulator and experiment control
SEP	somatosensory evoked potential
SEP	somatosensory evoked potential
sfMcGill	McGill pain questionnarie
SF-MPQ	short-form McGill pain questionnaire
Sias	best active site in each electrode
Sid	Selectivity index – device
SIDNE	stimulation-induced depression of neuronal excitability
SiNx	silicon nitride
SIROF	sputtered iridium oxide films
Slas	selectivity index
sLORETA	standardized Low Resolution Electromagnetic Tomography Algorithm
SMA	shape memory alloys
SNR	signal-to-noise ratio
SOM	somatostatin
SP	substance P
SVM	support vector machine
TA	tibialis anterior
TENS	transcutaneous electrical nerve stimulation
TEP	tactile evoked potential
tf-LIFE	thin-polymer-based electrodes longitudinally in the nerve
tf-LIFE	Thin-film Intrafascicular Multichannel electrode
TIME	Thin-film Intrafascicular Multichannel electrode
TIME	Transversal intrafascicular multichannel electrodes
TIME-3H	Transversal intrafascicular multichannel electrodes – human
TMR	targeted muscle reinnervation
TNF-α	tumor-necrosis factor-α
tr	training set
TRP	transient receptor potential
TRPV1	vanilloid receptor
USEA	Utah slanted electrode array
VAS	visual analog scale
VE	multi site stimulation
VPL	ventral posterior lateral
WDR	wide dynamic range neurons
ZIF	zero insertion force

Introduction

Winnie Jensen

Department of Health Science and Technology, Aalborg University, Denmark
E-mail: wj@hst.aau.dk

Amputation of a limb is a surgical intervention used as a last resort to remove irreparably damaged, diseased, or congenitally malformed limbs where retention of the limb is a threat to the well-being of the individual. The procedure traumatically alters the body image, but often leaves sensations that refer to the missing body part, the phantom limb. In 50–80% of cases, these sensations are perceived as painful and referred to as "phantom limb pain" (PLP). Today, it is still not completely understood why the pain occurs, and there are no effective treatments.

A Possible Path for Combatting PLP?

Cortical reorganization has been found to be related to PLP. Amputation of a hand is immediately followed by significant reorganization in the somatosensory pathway and cortex, i.e. the hand area in the brain is invaded by neighboring areas, such that the normal homunculus is shifted. Painful sensations appear to be related to reorganization of the primary somatosensory cortex (S1), and a correlation was demonstrated between the number of sites in the stump from where stimuli evoked referred sensations, the PLP experienced and the amount of cortical reorganization (Grüsser et al., 2001; Knecht et al., 1996). Several studies have demonstrated the favorable effect of enhancing the sensory feedback related to the missed limb to alleviate PLP in the recent years. For example, patients with PLP, who intensively used myoelectric prosthesis (Lotze et al., 1999) or used daily discrimination training of surface electrical stimuli applied to the stump

Hypothesis: By providing adequate patterns of stimulation to the transected afferent nerves central reorganization may be restored, and a normal processing of sensory signals recovered

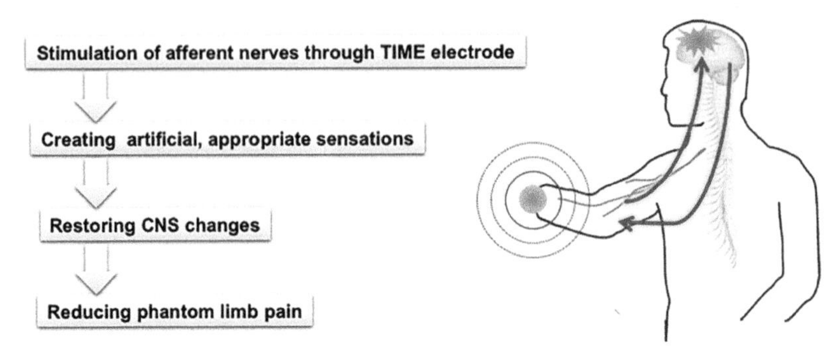

Figure 1 Overview of the main hypothesis of the TIME project.

experienced significant reduction of PLP (Floor et al., 2001). Intrafascicular, electrical stimulation of severed nerves proved to be capable of eliciting tactile or proprioceptive sensations by implanted LIFE electrodes in human subjects (Rossini et al., 2010). Rossini et al. also demonstrated that training for control of a robotic hand (with a limited amount of sensory feedback) significantly reduced PLP in a human amputee volunteer implanted with four LIFE. The reduction in PLP lasted several weeks after the LIFE electrodes were removed and changes in sensorimotor cortex topography were shown electrodes (Rossini et al., 2010). We therefore hypothesized that given appropriate control over a sufficient number of nerve fibers, a neural interface may be able to artificially evoke sensations and eventually relieve PLP (see Figure 1).

Why is a HMI-System is Needed to "Solve" the Problem of PLP?

As anybody who has felt a static electric discharge can attest, electrical impulses can be used to evoke sensations. These sensations are a result of the direct but nonspecific activation of peripheral nerves by the electrical discharge. If this basic principle is refined to specifically and focally activate only those nerve fibers projecting to sensory fibers related to pressure on the skin on the side of the index finger, for example, the evoked sensation would be that of pressure on the side of the hand. Given sufficient control over a large enough set of nerve fibers and types, the neural interface would be able to artificially evoke sensations of touch, vibration, heat, etc., and illusions of limb/finger/joint movement. The holy grail of human-machine interfaces has been considered a device that can directly interface to the body's nervous system. It has been the topic of popular science fiction but is based upon current experimental research in neuroprostheses. It is considered important because almost all interactions between the brain, the body, and the environment are relayed through information flowing through the nervous system. The ability to intercept information from, or artificially place the information into the nervous system can revolutionize the way the brain interacts with the body and the environment. But more importantly, such a technique may provide currently nonexistent treatment modalities to those who have lost or have pathological function due to traumatic injury or disease. Because of the sizes of the cells and constraints on dimensions of devices to minimize tissue damage, the only way to obtain high-density multichannel interfaces to the nervous system is through the application of micro/nanotechnologies to this medical device problem.

The TIME Prototype System for Treatment of PLP

Our aim was to develop such a Human Machine Interface (HMI) by means of the application of multichannel microstimulation to the nerve stump of an amputee volunteer to manipulate his/her the phantom limb sensations and explore the possibility of using the method as a treatment for clinched fist phantom limb pain.

This book provides an overview of our experiences and results with the design, development, and test of the hardware and software components, and our ambition to safely implant and evaluate the system in an amputee volunteer subject (see Figure 2).

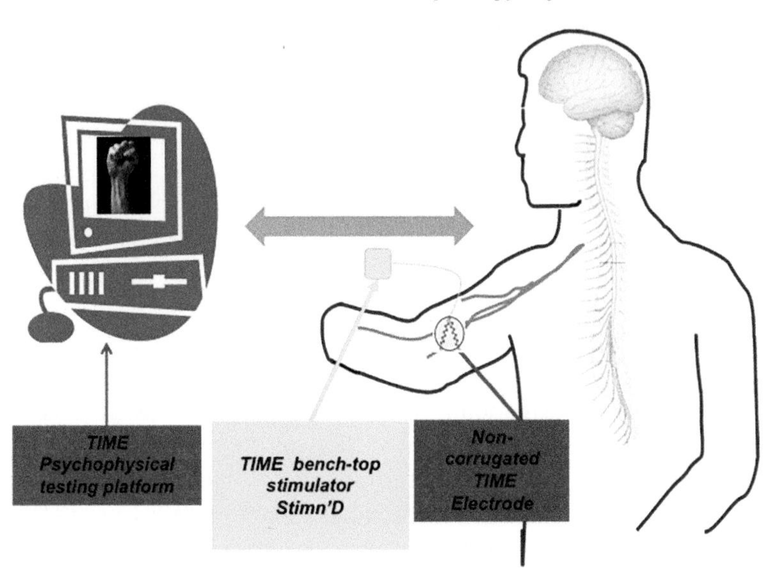

Figure 2 Overview of the TIME prototype system for preclinical evaluation in amputee subjects.

The "TIME prototype system" (see Figure 3) consists of the Thin-film Intrafascicular Multichannel Electrodes (TIME) (Chapter 3), a multichannel stimulator system (Chapter 7), and a psychophysical testing platform (Chapter 8). Theoretical modeling was carried out to drive electrode design (Chapter 4). The TIME electrodes underwent in vivo characterization in animals to test the biocompatibility, stability, and chronic safety (Chapters 5 and 6) before the system was tested in one human volunteer subject (Chapter 9). We also speculate on the future of the TIME electrodes and TIME prototype system (Chapter 10). Finally, to provide the reader with a broad background, we introduce the pathophysiology of pain (Chapter 2) and provide an overview of the current understanding and treatment of phantom limb pain (Chapter 1).

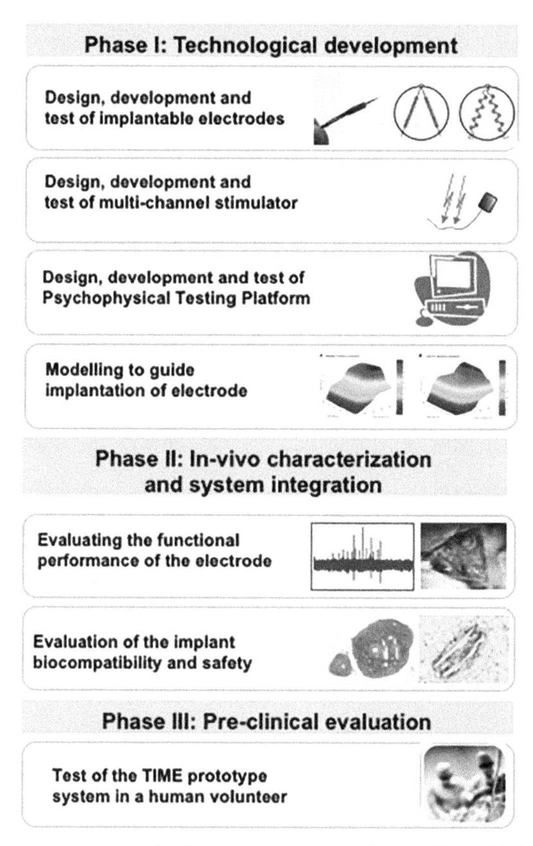

Figure 3 Overview of elements in the development and test of the TIME prototype system for preclinical evaluation in amputee subjects.

References

Flor, H., Dencke, C., Schaefer, M. and Grusser, S. M. (2001). Effect of sensory discrimination traning on cortical reorganization and phantom limb pain. Lancet. 357:1763–4.

Grüsser, S. M., Mühlnickel, W. and Schaefer, M. (2004). Remote activation of referred phantom sensation and cortical reorganization in human upper extremity amputees. Exp Brain Res. 54:97–102.

Knecht, S., Henningsen, H., Elbert, T., Flor, H., Höhling, C., Pantev, C. and Taub, E. (1996). Reorganizational and perceptual changes after amputation. Brain. 119:1213–1219.

Lotze, M., Grodd, W., Birbaumer, N., Erb, M., Huse, E. and Flor, H. (1999). Does use of myoelectric prosthesis reduce cortical reorganization and phantom limb pain? Nature Neurosci. 2:501–2.

Rossini, P. M., Micera, S. and Benvenuto, A. et al. (2010). Double nerve intraneural interface implant on a human amputee for robotic hand control. Clin Neurophysiol. 121:777–83.

1

An Introduction to Phantom Limb Pain

Caleb C. Comoglio[1], Kristine Mosier[2] and Ken Yoshida[1,*]

[1]Department of Biomedical Engineering, Indiana University – Purdue University Indianapolis, Indianapolis, Indiana 46202, USA
[2]Department of Radiology and Imaging Sciences, Indiana University School of Medicine, Indianapolis, Indiana 46202, USA
E-mail: yoshidak@iupui.edu
*Corresponding Author

With amputation comes many new experiences and sensations. Most credit the discovery and early characterization of phenomena associated with amputation to Ambroise Paré (16th century) and, nearly 250 years later, Silas Weir Mitchell in 1866 (Finger and Hustwit, 2003; Kline, 2016). Since then, substantial research has been conducted to further understand the consequences, mechanisms, and phenomena associated with amputation through the investigation of physical and psychological changes after amputation. This chapter has several goals. The first is to introduce the topic of amputation and the associated sequelae. Second, discuss the epidemiology and several proposed etiologies of the sequelae, focusing on phantom limb pain (PLP). Third, review methods for measuring the manifestation of PLP, specifically with respect to psychophysical aspects and cortical representation. Fourth, explore the proposed treatments of PLP and consider a potential new therapy paradigm.

Multiple studies have estimated the prevalence of limb loss and the subsequent effects of amputation. As many as 185,000 amputations occur every year in the United States (Owings and Kozak, 1998; Ziegler-Graham et al., 2008). It was estimated that 1.6 million Americans were living with the loss of a limb in 2005, which translates to a ratio of 1:190 Americans; 65% of these individuals have lower extremity amputations (Ziegler-Graham et al., 2008). Fifty-four percent of amputation cases occur after diagnosis of dysvascular disease and 70% of amputees with dysvascular disease (or 38% of the amputee population) were noted to have a comorbidity of diabetes (Ziegler-Graham et al., 2008). An unfortunate reality for many amputees is a relatively high rate of reamputation (26% among those with dysvascular amputation (Dillingham et al., 2005; Ziegler-Graham et al., 2008)). Reamputation refers to those who underwent an additional procedure or additional procedures to the previously amputated limb or the contralateral limb within 12 months of the original procedure. In 1996, US medical care costs exceeded $4 billion yearly for dysvascular amputations alone (Dillingham et al., 2005), which is only about half (54%) of the amputee community (Ziegler-Graham, 2008). Ziegler-Graham et al. predict the number of amputees in the United States will reach beyond 3 million by the year 2050. This, coupled with the high prevalence of postamputation pain (PAP) and the high degree of pain experienced, easily makes the case that phantom pain is a relevant problem. To further complicate the issue, the amputee community is ill-informed in regards to PLP; 41.6% of amputees have never heard of the phenomenon (Kern et al., 2012).

1.1 Epidemiology and Etiology of Phenomena and Sequelae Associated with Amputation

Individuals commonly notice the presence of a phantom limb shortly after amputation. This phenomenon, known as phantom limb sensation (PLS), is the mental construction of the limb that is no longer present postamputation. The phantom limb, or phantom, can be represented in a number of forms, from normal orientations to those that are not easily described or even physically possible. The phantom can also present pain to the amputee in many varieties, such as tingling, burning, stabbing, etc. This phenomenon is known as PLP or phantom pain. PLP is a subset of PLS where the sensations specifically cause discomfort. Amputees also experience other common painful phenomena, such as neuropathic pain (NP) and residual limb pain (RLP; also known as stump pain). NP is pain due to the damage or dysfunction of the somatosensory nervous system and RLP is pain in the remaining portion of the amputated limb. All of these painful phenomena fall under the umbrella of PAP.

1.1.1 Phantom Limb Sensation (PLS)

While the mechanism of the PLS phenomenon is not clear, it is common among amputees; as many as 80–90% of amputees experience PLS (Jensen et al., 1983; Ehde et al., 2000; Casale et al., 2009). In arm or leg amputees, PLSs are generally localized to the distal region of the phantom, i.e., the hand, foot, fingers, or toes, and are typically not constant (Jensen et al., 1985). Rather, the sensations peak intermittently, sometimes on a monthly basis and sometimes several times a week (Ehde et al., 2000; Kooijman et al., 2000). Sensations can be provoked in various ways, such as stump movement, touching the stump, and urination (Jensen et al., 1983). In a study involving 255 amputees, 79% reported nonpainful PLS, and of those individuals 27% (most common) described the sensations as tingling, 26% as itching, 13% as feeling asleep, among others (Ehde et al., 2000). Another related phenomenon is perceived movement of the phantom, where the amputee is able to consciously move the orientation or sense movement of the phantom. Eight days after amputation, 36% of amputees felt movement of the phantom with 19% feeling spontaneous movements (i.e., movements that were not consciously driven) (Jensen et al., 1983). Similarly, another study by Kooijman et al. found 38% to experience movement (Kooijman et al., 2000).

For some amputees, electromyogram (EMG) patterns in the stump during imagined movements of the phantom limb are distinguishable and nonrandom, indicating hand motor commands are preserved after amputation and there exists an inherent understanding of how to manipulate/move the phantom (Reilly et al., 2006). The modulation of signal seen in the stump did not appear in experiments with the intact limb, which supports current theories postamputation reorganization at some level.

The efforts to move the phantom were not only observed through muscle movements, but also through peripheral nerve activity, i.e., Dhillon et al. recognized nerve activity in the residual limb during attempted movements. Furthermore, they recognized activity in the central nervous system (CNS), specifically in the motor cortex, during phantom movements (Dhillon et al., 2004). These findings emphasize the current understanding of phenomena associated with amputation; the sensorimotor cortices and related peripheral innervation are actively involved in the perception of the phantom limb.

An altered kinesthesia is also common. For example, as many as 30% of amputees experience telescoping, which is the gradual shortening or retraction of the phantom limb, as depicted in Figure 1.1 (Jensen et al., 1983; Hill, 1999). In some amputees, the phantom limb no longer reflects the original anatomy. In this example, the phantom limb shortens and is drawn into the stump. In these situations, the residual limb and phantom hand or foot are no longer in an orientation that matches the original volume or limb, which causes confusion and concern to many amputees. Telescoping has also been linked to increased levels of phantom pain (Flor et al., 2006).

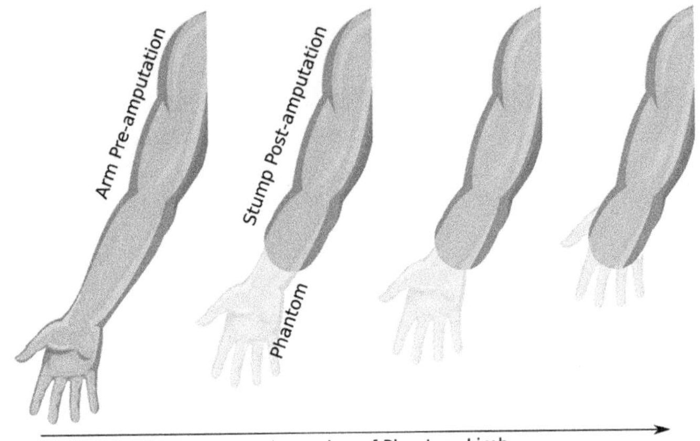

Figure 1.1 Among the peculiar potential pathological changes that occur after amputation is telescoping. Telescoping is a phenomenon in which the amputees sensory body image changes resulting in an alteration in the phantom sensations with respect to the sensations from the normal parts of the body. With time, the phantom sensation gradually moves or shrinks, for example as shown above, into odd or impossible positions or joint angles. This results in a state of sensory confusion, and concern for the amputee that potentially contributes to the increase in phantom limb pain.

In some circumstances, PLSs can be helpful in adjusting to the use of a prosthetic device, where the phantom limb embodies the prosthesis (Gallagher et al., 2008). Murray describes the embodiment phenomenon as a transition of a prosthesis from an extracorporeal structure to a corporeal one, meaning the prosthesis becomes part of the identity of self. This fits into the field of psychoprosthetics, which uses a psychological framework to analyze and explain the phenomena associated with prostheses and the amputation rehabilitation process. Corporeal embodiment does not occur in all amputees, which is not well understood. Murray attributes this embodiment transformation to practice, i.e., increased use of the prosthesis (Gallagher et al., 2008). Despite the possible utility of PLSs, in many cases the phantom sensation evolves into the form of PLP, which can be not only a hindrance, but also a phenomenon that has a strong negative effect on the amputee's quality of life (Knežević et al., 2015).

The phantom limb can also be debilitating when the sensations are painful; 54% of amputees who experience painful phantom sensations, or PLP, regarded the pain as somewhat bothersome (27% said extremely bothersome) (Ephraim et al., 2005).

Phantom sensations are not pathognomonic to amputation of a limb (Buonocore, 2015). In fact, studies have recognized phantom sensations in other sensory systems. Phantom eye syndrome has been found to affect as many as 51% of patients with orbital exenteration with 26% feeling pain (Roed Rasmussen et al., 2009). Phantom eye sensations most commonly came in the form of elementary visual hallucinations such as white light or colored light and were triggered by darkness, stress, and fatigue, among others (Roed Rasmussen et al., 2009).

Another argued case of phantom sensation is tinnitus, where individuals experience phantom auditory sensations, most commonly described as ringing in the ears, steady tones, or hissing. Tinnitus has been linked to hearing loss, i.e., up to 90% of cases are linked to hearing loss (Shore et al., 2016). Like PLS, tinnitus describes false perceptions; however, tinnitus is unique because it also occurs in individuals who are otherwise healthy. Sectioning of relevant cranial nerves has not proven successful for the treatment of tinnitus, lending to support the current proposed mechanism of maladaptive neural plasticity (House and Brackmann, 1981; Shore et al., 2016).

1.1.2 Phantom Limb Pain (PLP)

The prevalence of PLP, or phantom pain, widely varies in literature. A survey by Ephraim et al. (with 914 respondents), phantom pain was reported in 79.9% of amputees with 38.9% reporting the pain as severe ($\geq$7 on a 0–10 analog scale) (Ephraim et al., 2005). Ephriam et al. recognized no significant difference of the rates of phantom pain based on etiology, age, or level of amputation; they also noted that the rate of PLP for upper limb amputees was 83%, consistent with the rest of the study population. Eleven percent of the amputees in this study were upper limb (10% unilateral), leaving 89% as lower limb (79% unilateral). The mean pain intensity for phantom pain of all study participants was 5.5 $\pm$ 2.6 (Ephraim et al., 2005). Others have found prevalence rates ranging between 40% and 85% (Sherman and Sherman, 1983; Ehde et al., 2000; Kooijman et al., 2000; Schley et al., 2008; Kern et al., 2012). Various explanations have been offered for discrepancies in the prevalence, such as response rates and bias from choice of study population. However, the clear cause of the differences is not known. The range for PLP prevalence in amputees generally referenced in literature is 50–80%.

The quantification and description of PLP is important in understanding the effectiveness of treatment. From the standpoint of self-reporting scales, pain can be defined in terms of intensity, affect, quality, and location (Jensen and Karoly, 2010). Most research studies have opted to primarily measure intensity and bothersomeness using the visual analog scale (VAS) or the discrete version called the numeric rating scale (NRS). Average ratings of pain, in terms of the VAS, fall in the range of 5.1–5.5 out of 10 (Ehde et al., 2000; Ephraim et al., 2005). Ehde et al. found that when asked how bothersome the pain is (scale of 0–10, 0 being not at all bothersome, 10 being as bothersome as could be) 32% of respondents reported pain as being severely bothersome (≥7) and only 10% rated the PLP as not bothersome at all (Ehde et al., 2000). Likewise, Ephraim et al. found only 19% of respondents not to be bothered by the PLP they experienced (Ephraim et al., 2005). Amputees tend to describe PLP as knife-like (stabbing), sticking, burning, squeezing, etc. (Jensen et al., 1983; Jensen et al., 1985; Montoya et al., 1997).

A final metric or description of PLP is needed to quantify frequency and length-of-time of the pain. Efforts have been taken to define how often amputees felt PLP, and how long the pain was present. Amputees suffering from PLP experience the pain at different intervals; 31% report a frequency less than 1 episode per month, 14% a few times a day, and 7% have constant pain (Schley et al., 2008). Another study found 14%, 24%, and 24% for the same time frames, respectively (Kooijman et al., 2000). Kooijman et al., in the same work, found a fairly uniform distribution among frequencies of phantom pain attacks from feeling PLP a few times per year, month, week, day and constant pain, ranging from 14% to 24%. Kern et al. found of those experiencing PLP, 56.1% have pain lasting less than 5 h daily and many (27%) felt pain constantly (Kern et al., 2012). Ephraim et al. reported frequency in terms of never, sometimes, and always (20.1%, 58.7%, and 21.2%, respectively) (Ephraim et al., 2005). Ehde et al. found 81% of amputees to experience intermittent PLP, between once a week or less and four to six times per week (Ehde et al., 2000). Among these studies the rates are different for frequency of pain, as shown in Figure 1.2.

The median follow-up period for the study by Schley et al. was 3.2 years while the median follow-up period for the study by Kooijman et al. was 19.1 years. Also, the events leading to amputation (i.e., the study population) were slightly different among studies, where 98% of the Schley et al. data came from traumatic cases (Schley et al., 2008), 78% from traumatic cases in the study by Kooijman et al. (2000), and 50% for the study by Kern et al. (2012). Conversely, frequency and duration of PLP have also been found to decrease within 6 months after amputation (Jensen et al., 1985); this contradicts the discrepancy in the constant pain rate between Schley et al. (7% at 3.2 years after amputation) (Schley et al., 2008) and Kooijman et al. (24% at 19.1 years after amputation) (Kooijman et al., 2000). It is not clear which findings are

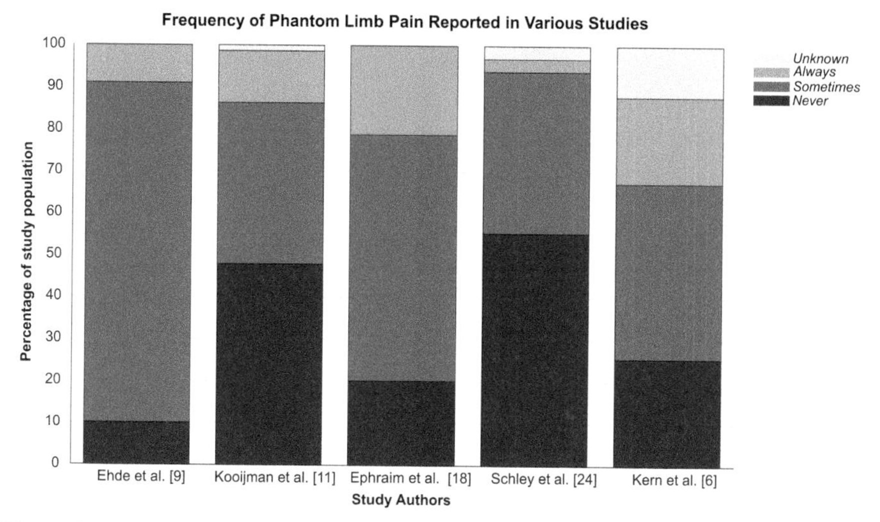

Figure 1.2 Various rates have been reported in literature for the frequency of PLP episodes. Most respondents reported PLP as occurring at a frequency somewhere between never and always. Several variables could explain discrepancies among studies, including epidemiology and etiology of amputation, years since amputation, size of sample population, etc. The effect of these factors on PLP presentation is not well understood.

more representative of the general amputee population. Ephraim et al. found of amputees 10+ years postamputation; 74% were experiencing phantom pain (Ephraim et al., 2005). The measure of length-of-time of pain has been reported in several ways, which makes it difficult to compare among reports in literature. PLP tends to flare episodically for seconds to minutes, but some have reported pain lasting several hours to a day or even longer (Jensen et al., 1985; Montoya et al., 1997; Ehde et al., 2000).

1.1.2.1 Triggers of PLP

Some have sought to understand the common comorbidities and triggers associated with phantom pain. Those who indicate a depressed mood are more likely to report severe pain and pain that is extremely bothersome (Ephraim et al., 2005). Phantom pain comes in many forms with many triggers. Often times PLP can flare during emotional distress, stump pressure, urination, cold temperature, or while coughing (Jensen et al., 1983). Preamputation pain has been recognized in several studies to be associated with phantom pain after the amputation (Jensen et al., 1983; Jensen et al., 1985; Schley et al., 2008). Many have suggested a correlation of PLP and RLP; however, Kooijman et al. suggested that RLP acts as a trigger of PLP (Kooijman et al., 2000).

This claim has not been substantiated by subsequent research. Giummarra et al. suggest several categories of triggers, the most frequent of which is “Movement and ‘behavioral schema’ triggers”; these include activities such as scratching an itch, gesturing with the phantom, etc. (Giummarra et al., 2011).

1.1.3 Residual Limb (stump) Pain (RLP)

A substantial number of amputees experience pain in their residual limb. As with other descriptors of pain, the rates vary widely in literature. Rates of stump pain span from 22% to 76% (Jensen et al., 1983; Smith et al., 1999; Kooijman et al., 2000; Ephraim et al., 2005; Schley et al., 2008; Bekrater-Bodmann et al., 2015). More recent surveys support rates on the higher side (61–67.7%) (Ephraim et al., 2005; Schley et al., 2008; Bekrater-Bodmann et al., 2015). Ehde et al. reported that, in response to asking which pain is the worst, the highest rated site (33%) was the residual limb, over phantom limb, back, and others (Ehde et al., 2000). RLP was also found in another study to be more impairing than PLP or back pain (Marshall et al., 2002). Only 4–13% of amputees experiencing RLP think of it as not bothersome at all (Ehde et al., 2000; Ephraim et al., 2005). On average, the intensity of the RLP falls in the moderate pain range at 5.4 on a 0–10 scale and is commonly described as aching or burning (Ehde et al., 2000). This is supported by Ephraim et al., who found that for the individuals experiencing RLP, the pain was almost uniformly spread among mild, moderate, and severe (41.8%, 28.3%, and 29.9%, respectively), with mild being slightly more prevalent (Ephraim et al., 2005). Similar to PLP, RLP tends to present itself in episodes and can last seconds, minutes, hours, or longer (Ehde et al., 2000). RLP does not tend to diminish with time after amputation (Ephraim et al., 2005). Looking for the possible cause behind the pain is an elusive question. O’Reilly et al. propose the pain is a result of neuromata (O’Reilly et al., 2013, 2016), which are sensitive bundles of nerve endings that result from inability to reconnect with the target tissue (Fried et al., 1991). Taken together, the high rate of prevalence and the impact on the quality of life highlight the degree to which RLP is a debilitating problem that needs to be addressed. A clear path to treating the issue of RLP is to look at treatment methods for NP. Neuromata are often associated with this type of pain, since inherently neuromata are a result of damage to the peripheral nervous system (PNS).

Amputees, often times, cannot distinguish between PLP and RLP (Hill, 1999; Flor, 2002). Generally, this confusion arises when pain is felt in the vicinity of the amputation site, where the phantom and residual limbs meet. RLP and PLP tend to correlate, especially in intensity (Ehde et al., 2000). Schley et al. found that 86% of amputees experiencing phantom pain also experienced stump pain (Schley et al., 2008).

1.1.4 Neuropathic Pain (NP)

NP plays a role in phantom phenomena (Buonocore, 2015). Casale et al. suggest that there is a significant link between neuromata and PLP (Casale et al., 2009). Neuromata make the surrounding area more sensitive to stimuli (mechanical, chemical, electrical), which explains correlations of pain and various triggers (e.g., touch, mood, stress, etc.) (Casale et al., 2009). Many of the descriptors of PLP and RLP reflect what would be expected of NP, i.e., burning, stabbing, etc., which leads one to conclude that PLP and RLP are forms of NP, and may link to the development of neuromata in the stump. Neuromata are the most common cause of pain in one study (O'Reilly et al., 2016). However, not all neuromata result in pain. For example, the same study found 159 neuromata in the sample population, but only 91 (57%) were painful in response to transducer pressure (O'Reilly et al., 2013). Another study supports this finding with similar rate of pain occurrence at 67% (O'Reilly et al., 2016). Furthermore, neuroma excision is not always successful. In a small case study, neuroma excision relieved pain in only two of the six patients (Nikolajsen et al., 2010). On the other hand, retrospective studies of neuromata removal found surgery to be a very successful method for relinquishing pain (Ducic et al., 2008; Sehirlioglu et al., 2009). Nevertheless, even though the links among neuromata, PLP and RLP are uncertain, it does not rule out that PLP and RLP arise from NP origins. Nikolajsen et al. found a link of PLP to N-methyl D-aspartate (NMDA) receptors through treatment with ketamine and concluded that PLP and RLP have mechanisms linked to both peripheral and central systems (Nikolajsen et al., 1996).

NMDA is an excitatory neurotransmitter which interacts with NMDA receptors. NMDA receptors are known to be associated with neural plasticity, having a role in long-term potentiation and long-term synaptic depression. They are also involved in sensory transmission; A-delta and C fibers use NMDA receptors among others in transmitting painful stimuli up nociceptive pathways at synapses in the Rexed laminae of the dorsal horn (Bleakman et al., 2006). Furthermore, having these roles gives way to one of the current, proposed mechanisms for NP, which points to NMDA receptors as a culprit for injury-induced central sensitization leading to secondary pain presentations such as allodynia and hyperalgesia (Bleakman et al., 2006; Collins et al., 2010). For this reason, as discussed later, NMDA receptors are a popular target for medicinal treatment approaches to alleviate NP (Collins et al., 2010).

Whereas, PLP is pain in the phantom and RLP is pain in the stump, linking the two to NP offers an explanation that neither form of pain would exist without injury to the PNS. This also assumes that RLP and PLP are not generated through traditional means of activating nociceptor pathways. Although, this theory does not explain all observed conditions of phantom pain, e.g., people who are congenitally limb-deficient. As many as 20% of these individuals experience phantom limbs at some level (either sensation or pain), even though there is no injury, per se (Melzack et al., 1997).

1.1.5 Secondary Effects of PAP

It is not just the rate of amputations and the severity of the pain that makes this problem relevant, but also the impact of PAP on an individual's everyday life. The multifaceted attack of PAP through various mediums, such as PLP, RLP, and other forms, interferes with daily activities (Marshall et al., 2002). Amputation and PAP negatively affect the self-perceived quality of life through fatigue and diminished mood (Trevelyan et al., 2016). This leads to high rates of depression among amputees (as high as 41%) presenting a vicious cycle, as there are substantial links among depression, level of pain, and bothersomeness of pain for PLP and RLP (Cansever et al., 2003; Ephraim et al., 2005). Depression secondary to amputation could be remediated by educating the population on the risks of amputation and providing mental health services (Darnall et al., 2005).

1.2 The Proposed Loci and Mechanisms of PLP

1.2.1 Neurologic Locus of PLP

The root cause of PLP is not clear as effects of amputation appear in each level of the nervous system, indicating multiple compounding sources of pain. Evidence suggests that PLP is the result of a multifaceted, combined system response from cortical, peripheral, segmental, and even psychological origins (Flor et al., 2006). Most propositions of mechanisms discuss cause and effect on the level of the CNS or PNS. Because of the many proposed mechanisms, further partitioning is necessary. Therefore, mechanisms are discussed below according to the relevant neurologic locus: peripheral, spinal, supraspinal, and cortical (Flor et al., 2006; Hsu and Cohen, 2013).

1.2.2 Predominant Mechanisms of the Peripheral Neurologic Locus

The Tinel sign (also "tingling" sign) was originally proposed to identify regions of peripheral nerve regeneration, specifically regarding cases of nerve injury (Davis and Chung, 2004). Similarly, one can use the Tinel sign on an amputee to locate nerve injuries that cause sensations or pain in the stump or phantom (referred sensation or RS). Commonly, the location that causes sensation or pain is at the site of a severed nerve, which has morphed into a neuroma. These neuromata (known as terminal neuromata) are typically formed within 1–12 months after nerve transection (Boutin et al., 1998), but start to form within hours (Fried et al., 1991). A study in rats found that ectopic discharges from injured peripheral nerves have a role in initiating NP, but do not have a significant role in the maintenance of NP (Sun et al., 2005). The onset of ectopic discharges is correlated with the onset of allodynia (pain from a stimulus that would normally be nonpainful) shortly after nerve transection, indicating these are responses to or results of injury (Sun et al., 2005). However, in animal studies ectopic discharges diminished over time, while tactile allodynia was maintained (Sun et al., 2005; Flor et al., 2006). These circumstances in the periphery seem to demonstrate two effects of nerve transection, but do not identify the source or mechanism of pain. For example, neuromata have been found to be sensitive to mechanical and chemical stimuli (Fried et al., 1991; Flor et al., 2006), so much so that PLP can be heightened from tapping (Nystrom and Hagbarth, 1981). However, a study on two amputees found that PLP persisted even after blocking PLP associated neuromata with lidocaine (Nystrom and Hagbarth, 1981). This causes further suspicion that PLP and other phantom phenomena are not caused by peripheral mechanisms; rather, they are merely accentuated by peripheral factors.

1.2.3 Predominant Mechanisms of the Spinal Neurologic Locus

Deafferentation of the dorsal horn is thought to be linked to PAP, specifically through central sensitization, which is the increased activity of the dorsal horn afferent targets due to decreased suppression from the brainstem (Iacono et al., 1987; Hsu and Cohen, 2013).

Deafferentation could be a result of amputation, or it could be another type of injury such as brachial plexus injury. Jensen et al. proposed that pain may be induced from atrophy of deafferented dorsal horn neurons and changes to receptive fields in the spinal cord (Jensen et al., 1983).

Spinal reorganization has also been recognized in functionally inactive regions and is reversible if the relevant nerves regenerate (Devor and Wall, 1981; Hsu and Cohen, 2013). It has also been manipulated through operant conditioning of spinal reflexes (a well-known mechanism for learning). Thompson and Wolpaw reviewed several studies that took advantage of the operant conditioning paradigm to alter reflexes (Thompson and Wolpaw, 2014). Because of the integration of sensory information in the spinal cord (especially connections involved in gating through suppressive inhibitory interneurons), spinal mechanisms are important to consider (Teixeira et al., 2015).

1.2.4 Predominant Mechanisms of the Supraspinal Neurologic Locus

Florence and Kaas found in animal studies that cortical reorganization was linked to reinnervation and sprouting afferents subcortically in the brainstem and thalamus (Florence and Kaas, 1995). Some have linked amputation to significant changes to the cuneate nucleus in the brainstem, which typically projects to the thalamus and transmits afferent sensory information, especially from the hand (Florence and Kaas, 1995; Wu and Kaas, 2002). Xu and Wall found changes in the cuneate nucleus to occur within minutes to hours after injury in primates (Xu and Wall, 1997). Further evidence of supraspinal reorganization was demonstrated in adult squirrel monkeys (Churchill et al., 2001). Churchill et al. found that somatotopic reorganization of the thalamus and brainstem was of a similar extent to what is reported for the cortex (Churchill et al., 2001).

1.2.5 Predominant Mechanisms of the Cortical Neurologic Locus

A traditional theory, as proposed by Ramachandran et al., is that cortical reorganization is the primary mechanism of PLP, which is typically discussed in terms of plasticity of the primary somatosensory cortex (S1) (Ramachandran et al., 1992, Flor et al., 2000). Directly following amputation, the mapping of

S1, i.e., Penfield's Homunculus, no longer matches the anatomical structure. Changes occur in the sensory and motor cortices adapting to both the altered anatomy and the loss of sensory input (Flor et al., 2000). Specifically, the plasticity of the cortex allows neighboring regions of the somatosensory homunculus to take over the region that previously mapped to the, now deafferented, limb (Ramachandran et al., 1992). However, this mechanism also has missing links when looking at clinical experiences. A case study of two amputees found that some experience RS in the phantom hand while touching the ipsi- or contra-lateral foot (Grüsser et al., 2004). Another study found RSs in the upper leg and genitals that mapped to the phantom in upper limb amputees (Giummarra et al., 2011). Flor et al. found significant differences in activity among amputees experiencing phantom pain compared to those not experiencing PLP in regions such as SI, the secondary somatosensory cortex (S2), and the posterior parietal cortex (PPC) (Flor et al., 2000). Other cortical changes have also been evaluated, such as unmasking of preexisting synapses of neighboring cortical regions, e.g., of SI, and of preexisting transcommissural connections, e.g., for coordinated movements of multiple limbs (Giummarra et al., 2007). The latter is of particular interest because it may explain cortical reorganization ipsi-lateral to the amputation as seen by (Schwenkreis et al. 2003; Garry et al., 2005).

1.2.5.1 Referred sensation and related mechanisms

While all phantom sensations are in a sense "referred," the definitions of PLS and RS are slightly different. PLSs are generally understood to be any sensation felt in the phantom limb, whereas RSs are perceived feelings in a body part when another body part is being stimulated (such as the residual limb or the face). RS is a common occurrence in amputees (Ramachandran et al., 1992; Flor et al., 2000). While it is possible to feel RSs without nerve injury by stimulating proximal regions of a peripheral nerve as demonstrated by Forst et al. (2015), RSs typically are amplified in amputees (i.e., more regions of the body such as the face and ear map to the phantom limb). Similar to amputation, substantial RSs have been noted in individuals with type I complex regional pain syndrome (CRPS) (McCabe et al., 2003), spinal cord injury (Moore et al., 2000; Soler et al., 2010), and other nerve-related ailments. As with other aspects of phantom phenomena there is debate on the mechanism of RSs. This phenomenon is thought to originate from mechanisms that are separate from other phantom phenomena, as they are non-neuropathic in nature (Buonocore, 2015). Flor et al. found correlation of RSs to increased activity of the PPC (Flor et al., 2000), while Ramachandran et al. supported reorganization of S1 to be the primary mechanism (Ramachandran et al., 1992). Stimulation of the remaining nerve in the residual limb has also elicited RSs; Dhillon et al. achieved this through stimulation with implanted electrodes (Dhillon et al., 2004). Similarly, Forst et al. were able to evoke RSs through surface electrical stimulation in healthy subjects by placing surface electrodes over the ulnar and median nerves (Forst et al., 2015).

The mapping of RSs requires the analysis of three primary locations: (1) the area being stimulated, (2) the area being referred, and (3) the cortical location of somatosensory processing. Several questionnaires call for a subject to locate the areas of pain (Melzack, 1975), but because nonpainful sensation are generally not bothersome (Smith et al., 1999), the location and mapping of RSs have not been addressed except cortically. This is a useful measure to determine changes in the presentation of pain. RSs can be evoked by touch; the Tinel sign is a simple method for identifying these regions (Trotter and Davies, 1909; Davis and Chung, 2004).

Several interesting phenomena, which likely have different mechanisms, are considered RSs. For example, the RSs evoked by touching the face of an amputee (as done by Ramachandran et al.) likely has a mechanism primarily in the cortex (Ramachandran et al., 1992; Flor et al., 2000). However, an RS evoked from stimulation of the proximal region of a peripheral nerve (as done by Dhillon et al.) likely can be explained by peripheral and/or spinal mechanisms (Dhillon et al., 2004; Forst et al., 2015).

1.2.6 Psychological Aspects of Pain

Emotional and psychological states have a large role in interfering with amputees' lives (Shukla et al., 1982; Kashani et al., 1983; Cansever et al., 2003). The initiative on methods, measurement and pain assessment in clinical trials (IMMPACT) recommends testing effects on emotional functioning when conducting pain-related clinical trials (Dworkin et al., 2008). Since amputees have exhibited differences from the general population in this respect, it is reasonable to assume that it also plays a role in the experience of PLP and other postamputation phenomena. In general, PLP is not a symptom of psychological distress (Katz, 1992). Katz and Melzack reported that depression and anxiety were not predictors of PLP (Katz and Melzack, 1990). This is further supported by Darnall et al. who found extremely bothersome RLP or PLP lead to increased odds of depressed symptoms, but depressed symptoms do not necessarily indicate bothersome RLP or PLP. They concluded that one of the highest risk factors for depressive symptoms is PAP (Darnall et al., 2005). Both Hill and Katz cautioned researchers on the assumptions related to depression and PLP saying claims of psychological explanations of pain are unsubstantiated and study populations may be inherently biased (Katz, 1992; Hill, 1999). Along the same lines, some have suggested that the causal relationship between pain and mood is only unidirectional, i.e., negative mood states are a result of pain, but pain is not a result of negative mood (Blågestad et al., 2016). Even though the relationship of PAP and depression is still under investigation, the relationship of depression and amputation seems to be quite clear. In addition to depressive symptoms, evidence of anxiety, insomnia, and other psychological ailments are prevalent (Shukla et al., 1982). This demonstrates a need for mental health services among the amputee population.

1.3 "Phantom" Pain in Nonamputees – A Complicated Issue

The traditional definition of PLP refers to pain in a limb that is not present. However, there are also instances of sensation and pain in a limb that has lost connection to the CNS (deafferentation), from brachial plexus avulsion (BPA) or intraspinal injury, for example. These scenarios have been dubbed as "phantom" because the individual does not experience pain or even sensation through typical nociceptive and sensory pathways, because they are no longer connected. In this regard "phantom" sensations have been found in individuals who have brachial plexus injuries (Sweet, 1975; Son and Ha, 2015; Tsao et al., 2016). In addition to the similar descriptions of pain, after BPA individuals experience RSs in the deafferented limb from touching the ipsilateral face (Tsao et al., 2016). Brachial plexus injuries also lead to cortical reorganization (Qiu et al., 2014). Most often pain is described as tingling, pins and needles, burning, sharp, or paroxysmal (Parry, 1980), which is reason to believe BPA causes NP (Teixeira et al., 2015). The underlying mechanisms of pain as a result of BPA are not well defined. In comparing symptoms, one must consider that brachial plexus injuries are often incomplete, meaning the limb remains partially sensate because it is still partially neurologically intact. If individuals with BPA or intraspinal injury experience PLP, the phantom pain and phantom sensations convolute with trace sensations from the limb. Furthermore, the presence of the limb further complicates discriminating phenomena as phantom or not. While the pain presents in a similar fashion to that of pain as a result of amputation, the presence of the limb makes it difficult to know if the mechanisms are the same.

1.4 Theories of Why PLP Presents

In the study of phenomena associated with amputation, an important thought to consider is that a single mechanism will likely not explain all phenomena. This idea was proposed by Sherman et al. in their evaluation of the mechanism of PLP, which concludes that different presentations of pain should be treated differently clinically, but does not suggest how (Sherman et al., 1989). Several theories have been proposed over the years to explain PAP and phantom phenomena. Ronald Melzack and Patrick Wall have had many contributions to this list and evolution of theories including the Gate theory of pain, the Neuromatrix theory, and others, which are discussed further.

1.4.1 Gate Theory

Gate theory is a prominent pain theory developed in the 1960s (Melzack and Wall, 1965). The concept in its most basic form can be summarized as a complex multi-input, multilayered system, where inputs at various layers can relay "off" or "on" signals, which cascade to determine whether or not pain is perceived (Melzack and Wall, 1965; Mendell, 2014). More specifically, Gate theory suggests that portions of the dorsal horns, such as the substantia gelatinosa, and the brain are active contributors to the system, which excite, suppress, and modulate signals to downstream targets (Melzack, 1999). Wall reinforced the theory after a few years discussing new findings in the field and how they relate to the previously proposed theory (Wall, 1978). In development of the theory there were many unknowns as to how the theory was implemented physiologically. In returning to the topic Wall proposed that descending control involves the periaqueductal grey matter and nucleus raphe magnus (Wall, 1978).

The theory was proposed ahead of its time, pushing the field forward to better understand mechanisms of pain (Mendell, 2014). Since its introduction, Gate theory has evolved over several decades to account for new findings (Wall, 1978; Mendell, 2014). It provided the framework for future theories of mechanisms that incorporate the CNS and an individual's unique life experiences (Hill, 1999; Melzack, 1999). Melzack proposed a new theory as a derivative from Gate theory called the Neuromatrix theory, which emphasizes a sense of self in the perception of pain (Melzack, 1999).

1.4.2 Neuromatrix Theory

The Neuromatrix theory relies on the concept of a network of neurons that defines a genetically determined feeling of self (Melzack, 1990, 1992). The neuromatrix is thought to extend beyond the somatosensory areas of the cortex to the limbic and thalamocortical systems (Giummarra et al., 2007). Melzack proposed the neuromatrix could be molded by sensory input and is comprised of "thalamocortical and limbic loops," which cyclically process and synthesize input and output patterns. These patterns are what Melzack deemed the neurosignature, an individual's pattern of synaptic connections impressed on the neuromatrix (Melzack, 1990). An altered neurosignature, due to amputation, for example, would result in the experience of a phantom limb through sensations and possibly pain (Flor, 2002). The Neuromatrix theory considers sensory input and transmission on a "level of equal importance" as hormonal mechanisms of stress, meaning pain does not exist solely in a space of neural mechanisms, but also has psychological factors (Melzack, 1990). The diffuse nature of the theory, i.e., pain (or even phantom sensation) being the output of a large, complex psychophysical system, makes it difficult to isolate and test clinically (Hill, 1999; Flor, 2002; Giummarra et al., 2007). Furthermore and even more perplexing, the theory does not offer an explanation for why some amputees experience phantom pain or phantom sensation and others do not (Flor, 2002). Giummarra et al. offer examples of seven phantom limb-related experiences that are not explained by the Neuromatrix theory and concludes that Neuromatrix theory may provide explanations of PLP, but not PLS (Giummarra et al., 2007). While Neuromatrix theory is intriguing and will likely spark discovery in the current age of pain research (like Gate theory did in the 1960s), it lacks some explanation for phantom phenomena.

1.4.3 Maladaptive Cortical Plasticity

The idea of maladaptive cortical plasticity is that the sensorimotor cortex reorganizes in a way that causes pain post deafferentation. Whereas, it is clear that the cortex reorganizes postamputation, the extent of the relationship between reorganization and pain is unclear (Flor et al., 2006). Evidence supporting this theory compared hand and lip movements among upper limb amputees and healthy controls, where amputees experiencing PLP showed reorganization of the mouth and hand region of S1 and the primary motor cortex (M1) (Lotze et al., 2001). In a study of brain-machine interfaces with patients experiencing phantom pain, Yanagisawa et al. found that attempting to merge and amplify neural signaling to cortical representation of the phantom actually increased pain (Yanagisawa et al., 2016).

1.4.4 Pain Memory

The pain memory hypothesis supposes that phantom pain mimics preamputation pain because of implicit pain "memories" established in the somatosensory system (Flor, 2002; Flor et al., 2006). The hypothesis relies on plasticity of the somatosensory cortex due to nociception (Flor et al., 2006). In a small study involving capsaicin injection, sensitivity of SI to nociception has been measured, improving validity of the hypothesis (Sörös et al., 2001). Further support for the hypothesis is that phantom pain commonly embodies pain that was experienced preamputation (Katz and Melzack, 1990), and several studies have found correlations between preamputation pain and phantom pain (Jensen et al., 1983; Nikolajsen et al., 1997). However, this theory does not account for the amputees who experience PLP but do not experience pain preamputation. Furthermore, some amputees feel pain due to the phantom limb being in an unnatural or biologically impossible orientation, which does not support this hypothesis.

1.4.5 Sensory Confusion

The hypothesis of sensory confusion assumes that pain is a result of ramping due to broken feedback mechanisms. While feedback loops exists subcortically, evidence also points to involvement of frontal and parietal brain areas in the "incongruence of motor intention and sensory feedback" (Harris, 1999; Flor et al., 2006). Similar to Gate theory, this hypothesis relies on closed-loop control of peripheral and central mechanisms, which modulate sensorimotor information during movement. Harris compares this effect to the feeling of nausea when senses do not agree on body position or balance (Harris, 1999).

1.5 Measuring PLP

Pain has both behavioral and physical properties and can be largely subjective. Intensity, affect, quality, and location are the primary experiential dimensions of pain (Jensen and Karoly, 2010). Pain intensity refers to the extent of the pain and can be subjective based on historical experience of the individual reporting the pain. Pain affect refers to the "emotional arousal or changes in action readiness caused by the sensory experience of pain," as so eloquently put by Jensen and Karoly (Jensen and Karoly, 2010). In essence, pain intensity refers to the extent of pain while pain affect refers to the emotional experience related to pain or the extent to which the individual is bothered by the pain. Pain quality refers to the descriptors of pain with respect to sensation, such as tingling, burning, sharpness, etc. and also includes the time-related aspects of pain, such as frequency, length-of-time of pain, etc. Pain location defines the area pain is perceived. Each of these four dimensions of pain is important to measure when studying the effectiveness of treatments and therapies for PLP. However, the measurement of PLP is a complicated issue. When measuring pain in a research setting (clinical or animal), there are additional considerations, such as the effects of habituation and sensitization (Johnson, 2016). Because of these barriers, pain-researchers utilize multiple measures and consider behavioral presentations of discomfort in analysis (Huskisson, 1974). Across studies of proposed therapy methods, various pain measures and scales have been utilized; in regards to PLP, studies tend to describe the degree of pain and the extent the pain interferes with the individual's life through various psychophysical measurement modalities (Hill, 1999). This variety of methods makes comparisons of results difficult.

1.5.1 Psychophysical Measures of Pain

In order to understand the effects of a given therapy modality, one must measure the various aspects of pain. Several validated measures are available to do this. The instruments used in the present study for effect determination are the VAS, neuropathic pain symptom inventory (NPSI), profile of mood states-short form (POMS-SF) and are discussed further.

1.5.1.1 Self-report questionnaire

While self-report questionnaires are an obvious way to gather information and understand the pain being perceived, the subject-to-subject (intersubject) variation cannot be predicted. For example, Dar et al. found, in a small study of injured veterans, that severely injured individuals have a higher pain tolerance and higher pain threshold than lightly injured individuals (Dar et al., 1995). In a study of thermal pain thresholds, Wasner et al. explored preconditioning as a means of testing sources of intersubject variations; however, in terms of pain thresholds, the study found no difference in subjects who were preconditioned and subjects who were not preconditioned (Wasner and Brock, 2008). This is a relevant finding because of the concern for scale recalibration presenting a potential source of variability in self-report data. The proposition of scale recalibration is an issue that is not addressed in the realm of PLP. However, in other research areas, this has not been validated as a source of variation. Lacey et al. found no evidence of scale recalibration in individuals suffering from chronic illness (specifically with regards to quality of life ratings) (Lacey et al., 2008). Nevertheless, studies typically rely on validated instruments and assessments to characterize pain and understand the effects of a given treatment for a population.

1.5.1.2 The visual analog scale (VAS)

Psychophysical measures involve those that describe an individual's perception. A commonly used instrument is the VAS. With respect to pain intensity, an individual experiencing pain ranks the pain somewhere between "no pain" and the "pain as bad as it could be" by marking a line spanning between the two extremes (commonly separated by 10-cm). The individual's severity of pain can be enumerated by measuring the length from 0 (no pain) to the marking (Huskisson, 1974). The primary measure of most studies describing the prevalence of PLP is typically some version of pain intensity; most often this is done with the VAS (Hill, 1999). The VAS and the discrete version, NRS, can be used for any measure in which there are two extremes. The VAS has been used to understand other aspects of phantom phenomena, such as intensity of PLS (Sherman and Sherman, 1983), and it can be useful in describing the effect of a treatment or therapy. In fact, it is used frequently outside of the realm of PLP (Huskisson, 1974). When describing the intensity of phantom pain, the VAS is often used along with the interpretation or adaptation into mild, moderate, and severe pain. Jensen et al. attempted to standardize these descriptors to pain ranges, 1–4, 5–6, and 7–10, respectively, by considering factors such as pain interference and impact on quality of life (Jensen et al., 2001).

1.5.1.3 The neuropathic pain symptom inventory (NPSI)

The idea of using a VAS or NRS has been adopted and adapted to quantify other unmeasureables because of its dependability (Huskisson, 1974). The NPSI utilizes several NRSs to quantify the qualities of NP (Bouhassira et al., 2004). Ultimately, the responses are combined to form subscores, which represent different aspects of NP, i.e., burning, pressing, paroxysmal, evoked, and paresthesia (or dysesthesia), and overall NP. In the case of NPSI, paresthesia/dysesthesia are defined by the same subscore, which is related to feeling pins and needles and feeling tingling (Bouhassira et al., 2004).

The usefulness of the NPSI is that it not only demonstrates the presence of NP, but also the presentation of the pain. Having this capability offers the opportunity to study the effects of treatment on subtypes of NP as well as the effects on overall NP. Mackey et al. proposed extracting information on NP from the short-form McGill pain questionnaire (SF-MPQ; discussed further in subsection\ref{opm}); this method takes advantage of an existing questionnaire, but it is not as specific as other measures, such as NPSI (Mackey et al., 2012). Other measures specifically related to NP exist, such as the neuropathic pain scale (NPS) (Galer and Jensen, 1997), the neuropathic pain questionnaire (NPQ) (Krause and Backonja, 2003), the "neuropathic pain four questions" (DN4) (Bouhassira et al., 2005), the Leeds assessment of neuropathic symptoms and signs (LANSS) (Bennett, 2001), among others; however, these alternative instruments are either not strongly validated, not detailed enough, or are designed to differentiate non-NP from NP and not to assess NP (Bouhassira et al., 2004). The NPSI has been validated in several languages among various populations (Bouhassira et al., 2004; Sommer, 2011; Matsubayashi et al., 2015). A German study found NPSI test-retest reliability to be suboptimal (Sommer, 2011), compared to the original study (Bouhassira et al., 2004). Although, in the German study the time lag was 24 h (compared to 3 h in the original study (Bouhassira et al., 2004)). While this is a notable finding, it does not change the validation of the instrument as it is reasonable to expect changes in the presentation of pain in a 24 h period; temporal variation is a known characteristic of NP (Gilron et al., 2006).

1.5.1.4 The profile of mood states-short form (POMS-SF)

In traumatic lower limb amputees, the prevalence of depression was 41.6% (Cansever et al., 2003). In a broader population base of various etiologies, significant depressive symptoms were seen in 28.7% (Darnall et al., 2005) (compared to 4.9% point prevalence) and 17.1% life-time prevalence in the general population (Blazer et al., 1994).

Ephraim et al. aptly noted the correlation of depression and the presence of PLP, where increased pain intensity corresponded to heightened depressive symptoms (Ephraim et al., 2005). The finding suggests that there is a need to continuously monitor and swiftly treat depression in amputees (Ephraim et al., 2005). In a more general sense, mood correlates to the intensity and perception of pain greatly (Blågestad et al., 2016). Some attempts have been made to treat pain using the class of drugs called antidepressants and through psychological treatments of pain Gilron et al., 2006; Alviar et al., 2016; however, these have been ineffective (Eccleston et al., 2015). Mood does not act as an effective target for treatment. However, it may act as an indicator of positive or negative effect because of its correlation to pain.

The POMS-SF is comprised of 37 descriptors of mood. Each descriptor is ranked by the study subject on a five-point scale (1 = "Not at all," 5 = "Extremely") and is incorporated into a subscale, which can be used to characterize the individual's mood. The subscales are depression, vigor, confusion, tension, anger, and fatigue. Whereas depression has been shown to positively correlate with pain, other mood descriptors could provide more insight on the relationship of PLP and psychological state.

1.5.1.5 The brief pain inventory-interference scale (BPI-IS)

The brief pain inventory (BPI) has been adapted into a more succinct questionnaire as the BPI-short form (BPI-SF), which is a validated instrument for pain interference (Tan et al., 2004; Osborne et al., 2006; Raichle et al., 2006). The final series of questions is known as the BPI-IS. Questions are nonspecific to phantom pain and describe how pain has interfered with daily living over the past 24 h. The seven-question interference scale utilizes 11-item NRSs to describe pain's interference with general activity, mood, walking ability, normal work, relationships with other people, sleep, and enjoyment of life. The NRSs span from 0 ("Does not interfere") to 10 ("Completely interferes").

1.5.1.6 Problems with measuring PLP and other phantom phenomena

One factor not addressed by Jensen et al. when describing the standardization of the VAS with respect to PLP is the associated anchors of the VAS (Jensen et al., 2001). Anchors are defined as the descriptions of the minimum and maximum scores. Jensen et al. used a scale of 0–10 with anchors of "0 = no pain" and "10 = pain as bad as it could be" (Jensen et al., 2001). A prime example of this inconsistency in research related to PLP can be found in reports of the intensity of pain. In Table 1.1, several examples demonstrate how intensities are reported among various authors. The outcome of not utilizing a standard instrument for measuring pain intensity is data that are not directly comparable. While it may be possible to normalize the various scales back to the standard scale proposed by Jensen et al., correlations have not been proposed among the various scales.

Furthermore, interpretation of changing VAS scores is nontrivial. Jensen et al. suggest that a change in pain intensity from "7 to a 4 might be considered more beneficial and more clinically relevant than a reduction from a 4 to a 1,

Table 1.1 Different investigators use VAS pain scales that quantify pain intensities using different anchors, making it difficult to compare the measures between studies

Reference	Pain Scale	Anchors
Sherman and Sherman (1983)	0–100	Anchors not described
Montoya et al. (1997)	0–10	No pain / Unbearable pain
Smith et al. (1999)	0–100	Extremely mild / Extremely intense
Ehde et al. (2000)	0–10	No pain / Pain as bad as it could be
Marshall et al. (2002)	0–10	No pain / Pain as bad as it could be
Ephraim et al. (2005)	1–10	Mild pain / Extremely intense pain
Schley et al. (2008)	0–100	Anchors not described

at least in terms of the impact of the treatment on function and quality of life" (Jensen et al., 2001). This conclusion suggests that both the change in pain intensity as well as the baseline or reference pain intensity are important factors to keep track of in establishing effective treatments and therapies.

1.5.2 Other Proposed Self-report Measures of PLP

Because of the lack of standardization, several questionnaires and instruments have been developed or adapted for measuring PLP. Hill notes in a literature review of PLP, the MPQ and SF-MPQ have been used in several studies (Hill, 1999). The MPQ and its variants have significantly contributed to the understanding of pain (in general) and PLP, and it acts as a primary instrument in many pain studies (Katz and Melzack, 2010).

Alternate measures of depression include the Center for Epidemiological Studies-depression questionnaire (CES-D) (Ephraim et al., 2005). The chronic pain grade (CPG) (Von Korff et al., 1992; Ehde et al., 2000; Marshall et al., 2002) distributes an individual's pain into one of four grades based on intensity and disability associated with pain. Grade I is the least intense and least disabling, while Grade IV is the most intense and most disabling (Von Korff et al., 1992). Flor et al. and Montoya et al. used a 122-item phantom-and-stump phenomena interview as a primary instrument (Flor et al., 1995; Ehde et al., 2000). The interview is a compilation of several standard instruments to separately analyze stump and phantom sensations and pain, including a modified version of the MPQ, several VASs to describe average pain severity and intensity of nonpainful sensations, descriptors of sensations, along with several open-ended questions (Ehde et al., 2000). Montoya et al. also utilized the West Haven-Yale multidimensional pain inventory (MPI) to evaluate the severity and interference of stump and phantom pain (Ehde et al., 2000). Smith et al. (1999) used the prosthesis evaluation questionnaire (PEQ;

developed by Legro et al. (1998)). The PEQ highlights intensity, frequency, and bothersomeness of phantom, stump, and back pain as well as phantom sensations (Smith et al., 1999). Further evidence of lack of standardization is that study designs have opted to utilize self-designed questionnaires such as the Groningen questionnaire problems after arm amputation (GQPAA) by Kooijman et al. (2000).

1.5.3 Measuring Cortical Reorganization

Cortical plasticity or cortical reorganization is a popular topic in the study of postamputation phenomena. This is mainly because of the desire to understand the underlying mechanisms. While plasticity is not unique to the cortex (Florence and Kaas, 1995), it gets particular attention because of the relationship of the somatosensory mapping and observations of RSs in the facial region (Ramachandran et al., 1992). From the perspective of characterization, studies have investigated the differences in cortical activity among amputees and healthy controls. Lotze et al. studied the locus of activation for hand and lip movements using functional magnetic resonance imaging (fMRI), comparing amputees with PLP ($n = 7$), amputees without PLP ($n = 7$) and healthy controls ($n = 7$) (Lotze et al., 2001). Reorganization of the hand and lip areas in M1 and S1 was recognized in patients with PLP but not others. Many studies have also investigated the cortical differences between the activities utilizing the affected limb versus the individual's healthy limbs. This paradigm attempts to have an individual serve as his or her own control. Measurement of changes to the cortex can be done through several modalities. Blood oxygen level dependent (BOLD) fMRI is used most often because of the ability to relate activation to particular cortical structures. Most studies that use event-related BOLD fMRI to look at cortical reorganization focus on S1 and M1 (Flor et al., 1995; Lotze et al., 2001). Other instruments include electroencephalogram (EEG) coupled with some type of somatosensory evoked potential (SEP) in the periphery, such as tactile evoked potential (TEP) or laser evoked potential (LEP) (Flor et al., 2000; Zhao et al., 2016). Coupling both EEG and MRI, Flor et al. used EEG to record cortical activation during RSs elicited by TEP, and used the activation map to overlay an anatomical image captured via magnetic resonance imaging (Flor et al., 2000).

Some disadvantages should be considered when using BOLD fMRI to study cortical differences. The main disadvantage is the length of time required for measurement. BOLD fMRI contrast relies on the hemodynamic response function (HRF), which is an increase in oxygenated blood (specifically oxyhemoglobin) compared to a resting state. The underlying assumption is that the increase in blood in a particular region is a causal, time-delayed effect of increased neuronal activity. These details reveal a reason behind the intensive time requirements of fMRI, as stimuli do not elicit instantaneous responses. Beyond the time dynamics of the biological system, the larger contributors to lengthy experimentation paradigms are issues of signal-to-noise ratio (SNR). To alleviate the poor SNR, fMRI paradigms typically utilize signal averaging, thus longer measurement times. Analysis of fMRI results involves an understanding of both estimation efficiency (ability to estimate the HRF) and detection power (ability to detect activation) as described by (Liu and Frank 2004; Liu, 2004). Furthermore, a recent study attempting to validate fMRI statistical analysis methods found high rates of false positives (Eklund et al., 2016).

1.5.4 Pros and Cons of Different Measurement Approaches

If relating back to the four primary dimensions of pain (intensity, affect, quality, and location), various instruments have positive aspects and points of weakness. For this reason, several research studies have implemented multiple instruments. Depending on the study design this could have different effects on self-report data. Thorough questionnaires and interviews (such as the MPQ or the phantom-and-stump phenomena interview) allow for detailed description of the pain, but take substantial time and concentration for the study participant. This could cause frustration and bias if the participant is enrolled in a study of temporal effects of treatment and having to complete a questionnaire multiple times, for example. Substantial effort should be taken to consider the length of time a study participant spends responding to questionnaires and the number of times a study participant responds to a particular questionnaire. On the other hand, there are disadvantages of being too brief (Jensen and Karoly, 2010). Brevity is just one consideration in the list of primary trade-offs, where targets should be set to reduce the required contact time between the health care provider (HCP) and patient, while maximizing the collection of relevant pain characterization data.

1.6 Current Treatment/Pain Management Methods

The proposition of treating PLP has been under study for decades. In 1980, Sherman et al. reported on 68 different possible methods (Sherman et al.,

1980; Sherman, 1980). To this day, a concise method for treatment has not been identified. Flor suggested more than 30 commonly used treatments for PLP in 2002, only a small fraction of which have shown any success in randomized controlled trials (RCTs) (Flor, 2002). Ideally, treatment methods of PAP and phantom limb phenomena would be developed from a mechanistic approach, i.e., the mechanism of pain would be utilized to address and reverse the pain. Since the mechanisms are not well understood, therapies tend to treat the symptoms, leading to a high number of available treatments, low rates of success, and high rates of dissatisfaction among patients (Sherman et al., 1980; Vernadakis et al., 2003). Current treatments of PAP can be broken down into medicinal and nonmedicinal methods. Medicinal treatments of pain utilize various methods of application: topical, oral, and local injection. A wide variety of nonmedicinal treatments have been explored, taking advantage of mechanical and electrical sensitivity of PAP. Other methods have used traditional pain management techniques, while some have ventured into the psychological treatment of pain. All-in-all treatment of any form of PAP has been largely unsuccessful.

1.6.1 Current Standard of Care

In 1983, a study found that only 17% of amputees were offered treatment for PLP even though 61% reported experiencing PLP (Sherman and Sherman, 1983). Several authors have noted a variety of responses from physicians to those suffering from PLP such as, "it is in your head" or PLP is "psychogenic" (Flor, 2002; Sherman et al., 1984; Mortimer et al., 2004; Sherman et al., 1987). Conversely, while the limb may no longer be present, the pain and sensations seem real. Another study in 1997 found nearly one-third of amputees who discussed PLP with their doctor were told no treatment was available (Wartan et al., 1987). Kern et al. attempted to study the success rates of relevant treatment methods by surveying amputees. Seventy-one percent ($N = 537$) of the amputees suffering from PLP had never received or sought after treatment; 19% felt their doctors were incompetent on the topic (Kern et al., 2012). Of those who did receive treatment for phantom pain, the treatment with the highest success rate was opioids via oral or IV administration at 67%. The second highest treatment method was opioid injection via intrathecal pump at 58%. Neither of these treat the root problem but only temporarily mask the pain (Kern et al., 2012). Whereas the medical and scientific communities are more accepting of the reality of PLP, the current standard of care is still up for debate. A focus group of health professionals found that information given to patients experiencing PLS and PLP is grossly inconsistent, indicating a necessity for a standard of care to be developed (Mortimer et al., 2004).

1.6.2 Medicinal Treatments

Medicinal treatments are among the most successful at alleviating PLP. Opioids/Opiates have shown a success rate as high as 67.4% (Kern et al., 2012), in particular morphine via injection and oral administration has shown successful reduction of but not elimination of PLP and RLP in a randomized controlled trial (Wu et al., 2002, 2008; Alviar et al., 2016). However, long-term analgesic efficacy has not been verified (Kern et al., 2012; Alviar et al., 2016). Anticonvulsants have also shown moderate success (52%) (Kern et al., 2012). Gabapentin is a commonly used anticonvulsant, which has had controversial results in RCTs. Bone et al. showed reduction of PLP in comparison to a placebo but no significant change in secondary measures, such as depression, mood, or sleep interference (Bone et al., 2002). Conversely, a separate RCT showed no significant difference between gabapentin and placebo groups (Smith et al., 2005). Some side effects were noted; however, these were not significantly different from the control groups (Bone et al., 2002; Smith et al., 2005).

Alviar et al. reviewed three NMDA receptor antagonists as possibilities: memantine, dextromethorphan, and ketamine (Alviar et al., 2016). The review identified only ketamine (Eichenberger et al., 2008) and dextromethorphan (Ben Abraham et al., 2003) to provide pain relief from this class of pharmacologic interventions (Alviar et al., 2016); however, both studies were underpowered (Alviar et al., 2016) and treatment with ketamine had substantial side effects, including dizziness, light hallucinations, and hearing impairment (Eichenberger et al., 2008). NMDA receptor antagonists have shown moderate success at relieving pain. The unsuccessful cases may be related to the mode of administration; each memantine trial reviewed utilized oral administration while other studies of this intervention method were successful with injection (Alviar et al., 2016).

Various other options have been explored and proposed for treatment including antidepressants, calcitonins, and local anesthetics (Alviar et al., 2016). In patient surveys, antidepressants have shown to be ineffective. Only 36.4% noted this method as effective (Kern et al., 2012). This ineffectiveness was supported in a RCT of amitriptyline that failed to show positive results (Robinson et al., 2004; Alviar et al., 2016). Furthermore, amitriptyline had a significant adverse effect of dry mouth over the placebo (Robinson et al., 2004). Local anesthesia was largely ineffective according to patient surveys (21.6% success) (Kern et al., 2012); RCTs of intravenous infusion with Lidocaine have shown successful treatment of RLP but not of PLP (Wu et al., 2002).

1.6.3 Nonmedicinal Treatments

Several nonpharmacological approaches have been proposed and tested as possible treatments for PLP, such as proper stump management, electrical stimulation, and mental imagery. Treatments vary significantly in regards to stimulus modality, psychological demand, and efficacy. Many therapies are proposed in case studies and uncontrolled trials, but either do not reach the stage of conducting a RCT or are not successful in a RCT, which makes identifying potential effective treatments in literature difficult (Halbert et al., 2002). Some of the more prominent methods are discussed further.

1.6.3.1 Nerve and stump management

Several methods have been proposed to thwart PAP related to neuromata; a universal method has not been accepted (Ducic et al., 2008; Vernadakis et al., 2003). Proper care of the stump and preventative measures in surgery are crucial to mediate pain. Painful neuromata are common among amputees; nearly 30% undergo surgery after amputation with the hopes of relieving neuroma-related pain (Kern et al., 2012). Often they form from improper surgical technique during the original amputation (Vernadakis et al., 2003). Studies have shown that simply excising the neuroma and applying traction to the nerve (encouraging the nerve to retreat into the stump) is not a successful procedure, only demonstrating successful results 33% of the time (Tupper and Booth, 1976). Over the years several techniques have emerged to ameliorate this painful phenomenon (Vernadakis et al., 2003). A recent review of neuromata treatment and prevention found nearly 200 techniques, supporting the perfect solution has not yet been found (Vernadakis et al., 2003). Some techniques have proven successful and appear notable; excision with silicone capping (83% success (Swanson et al., 1977)) or centrocentral anastomosis (94–95% success (Kon and Bloem, 1987; Barberá and Albert-Pampló, 1993)) are prime examples (Vernadakis et al., 2003). On the other hand, techniques such as these also present unnecessary risks to the patient. Silicone capping involves the introduction of a foreign body, which risks immunological response and inflammation in the stump (Ducic et al., 2008). Centrocentral anastomosis lengthens the time of surgery due to the meticulous nature of microsurgery, which means more opportunities for infection (Ducic et al., 2008).

One of the most notable techniques is nerve transposition (Vernadakis et al., 2003). Mackinnon et al. demonstrated the capability of minimizing neuroma formation in an animal model (Mackinnon et al., 1985). Rerouting the transected nerve into adjacent muscle without tension, resulted in significantly smaller neuromata compared to control groups in primate models (Mackinnon et al., 1985). Mackinnon and Dellon revisited the technique emphasizing the importance of separating the nerve ending from the scar tissue (Mackinnon and Dellon, 1987). This study found different success rate depending on a patient's previous experience ranging from 56% to 100% for good or excellent results (Mackinnon and Dellon, 1987). The nerve transposition technique had good or excellent results in 81% of cases (42 patients).

Another method that has had some success is targeted muscle reinnervation (TMR) (Souza et al., 2014). This is the act of intentionally ligating the original innervation of a nearby muscle to direct alternative peripheral nerves to the muscle. Generally, TMR utilizes a muscle that is no longer providing functional advantages to the patient with the hopes of the muscle acting as a target for the nerve. The long-term goal for these patients is that they could intuitively move their phantom, which would cause muscle activity in the targeted muscle; then, this muscle activation could be recorded, e.g., via EMG, to manipulate an active prosthetic. Conveniently, this method serves a dual purpose by also preventing the formation of neuromata. In a retrospective study 6 months after surgery, the method appears to be successful (Souza et al., 2014). All patients in this study reporting pain reported reduced or eliminated pain, and just under 90% were able to operate a TMR-controlled prosthesis.

Peripheral nerve surgery, such as TMR, is a treatment option for managing pain related to neuromata that has shown success in several studies, and is an excellent example of advancement in the field (Vernadakis et al., 2003); however, the degree of functionality provided by this method is often not necessary for lower extremity amputees. Rather than transferring the transected afferent nerve fibers to an alternative muscle or region, some have suggested merely tying the sensory nerves to nearby muscle away from areas forming scar tissue. If done during the amputation surgery, it could prevent formation and excision of the neuroma postamputation, thus lowering overall patient risk through reduction of procedures and procedural time (Ducic et al., 2008).

This procedure, proposed by Ducic et al. as an outpatient operation has had great success in a retrospective study of 21 neuroma excisions; patients reported an the average preoperative pain of 8.04 that decreased to 1.07 on the VAS (ranging 0–10) (Ducic et al., 2008). Furthermore, 85% reported improved quality of life. The key to this technique involved suturing the nerve-ending (after neuroma excision) to the nearby muscle. Some have proposed applying light traction to the nerve is sufficient, but an important detail to many of the techniques is to keep the nerve tension free (Vernadakis et al., 2003).

1.6.3.2 Electrical stimulation

Electrical stimulation of the residual limb, especially transcutaneous electrical nerve stimulation (TENS) or functional electrical stimulation (FES), has had success in case studies and small trials. However, as is the case with other therapy methods, the effectiveness of TENS has not been shown with a RCT (Johnson et al., 2015). Other forms of electrical stimulation have shown promise as well. Peripheral nerve stimulation showed significant improvement in regards to pain and quality of life, but the study lacked a placebo and had a small number of participants (Rauck et al., 2014). Others have attempted applying TENS to areas other than the residual limb, such as the contralateral limb (Tilak et al., 2016) and the ears (Katz and Melzack, 1991). Both of these methods showed a positive effect in small, short-term trials, but neither was compared to placebo groups. Sensory discrimination training using TENS has shown positive results (reduction in PLP and effect in cortical reorganization) in a small comparative study of 10 amputees (Flor et al., 2001). This method involved the application of random, nonmeaningful stimulation patterns of varying frequency, intensity, and location. Trial subjects were instructed to identify different patterns with the hypothesis that distraction from the pain actually reduces the pain (Flor et al., 2001). Success indicates there is a positive relationship among discrimination ability, cortical reorganization, and decreased PLP; although, the long-term effects of this method were not reported in (Flor et al., 2001).

1.6.3.2.1 *Considerations for FES of peripheral nerves*

Studying the effect in cats, Agnew et al. found that 8 h of high-rate, high-amplitude electrical stimulation resulted in irreversible damage of sciatic nerve axons (Agnew et al., 1999). In an earlier paper (McCreery et al., 1997), this effect was referred to as stimulation-induced depression of neuronal excitability (SIDNE). SIDNE, which according to the authors differs from long-term depression (LTD) because it does not involve a change in efficacy of the synapses and does not worsen day-to-day, can occur in the CNS if axons are subjected to "prolonged, high-frequency microstimulation" (McCreery et al., 1997). McCreery et al. stimulated the posteroventral cochlear nucleus (PVCN) for 7 h per day to find that with high enough intensity SIDNE could be induced, but was still reversible. The speculated mechanism attributed the effect to the entry of calcium into the neurons activating second messengers and several downstream pathways.

Lu et al. studied the effects of electrical stimulation on peripheral nerve regeneration in Sprague-Dawley rats (Lu et al., 2008). Methods involved transecting the right sciatic nerve, separating the nerve endings by 10-mm, and surrounding the nerve endings by a silicone rubber chamber. Stimulation was applied for 15 min every other day at 1 mA (1, 2, 20, 200 Hz depending on group). Results included histological samples as well as tests of nerve conductivity that showed the 2-Hz stimulation group to have the most mature structure. Lu et al. concluded that in regards to peripheral nerve regeneration, stimulation (depending on frequency) can have a positive or negative effect. Note, control group had 100% success in regenerating a nerve cable spanning the 10-mm gap; however, the conclusion was that the nerves generated under 2-Hz stimulation were healthiest (Lu et al., 2008). Cogan et al. suggest many culprits when it comes to the cause of tissue damage and that macroelectrodes and microelectrodes have different challenges when it comes to preventing tissue damage (especially charge density and charge per phase), but they did not address continuous stimulation (Cogan et al., 2016). Patel and Butera used stimulation frequency of up to 70 kHz to block nerves, but did not report on the possible effects of continuously stimulating at these high frequencies (Patel and Butera, 2015). Prodanov et al. (2003) reviewed FES in 2003 and pointed to two other articles by McCreery et al., which also discussed the negative effects of continuous electrical stimulation (McCreery et al., 1992, 1995). The 1995 McCreery paper indicates that low-frequency stimulation does not lead to early axonal degeneration, independent of stimulus amplitude.

1.6.3.3 Imagery

Mental imagery coupled with various techniques, such as muscle relaxation (Brunelli et al., 2015) or virtual visual feedback (Ramachandran and Rogers-Ramachandran, 1996; Mercier and Sirigu, 2009), present enlightening results that may reveal psychological aspects of PLP. Ipsi-lateral cortical reorganization could be a target for mental imagery, especially when utilizing coordinated bimanual movements through visual feedback (Schwenkreis et al., 2003; Garry et al., 2005). Mental imagery and muscle relaxation showed a significant reduction in PLP, PLS, and pain interference compared to a positive control group (Brunelli et al., 2015). The positive control group maintained the same physical therapy schedule as the test group, while the test group exercised mental imagery, in addition to the physical therapy. The success of this trial demonstrates an advantage of coupling physical stimulus with psychological exercise. Graded motor imagery (GMI) utilizes gradual training in three strategies: (1) implicit motor imagery, (2) explicit motor imagery, and (3) mirror visual feedback (Priganc and Stralka, 2011). Implicit motor imagery training involves laterality recognition or identification of images representing left limbs versus right limbs; explicit motor imagery practices movement of the phantom limb, or focusing on consciously manipulating the phantom; and, mirror visual feedback exercises the movement of the phantom while the patient utilizes visual feedback. Typically, the visual feedback involves placing the contralateral limb in front of the mirror, the amputated limb behind the mirror, and simultaneously moving both the contralateral and phantom limbs. Bowering et al. reviewed studies, including work on PLP by Moseley (2006), using this multipronged approach and found it to successfully treat chronic pain (Bowering et al., 2013). While the method has been proposed to treat PLP and PAP, the effects have not been thoroughly evaluated in this context (Limakatso et al., 2016). Some have compared the effects of mental imagery through virtual visual feedback (also known as mirror therapy) to that of TENS when applied to the nonamputated limb (Tilak et al., 2016). Both groups showed reduction in pain over a 4-day treatment phase, but neither group performed significantly better than the other.

This type of mental imagery could be considered a form of conditioning, where participants actively and consciously reinforce imagined movement with feedback (e.g., visual or tactile).

Imagery is supported by Macuga and Frey (2012), who found that imagery, i.e., actively simulating movements, stimulates more brain regions than passive observation. Studies on operant conditioning have shown to alter CNS organization in the spinal cord, specifically through retraining of spinal-cord-mediated reflexes (Thompson and Wolpaw, 2014). Thus, in these circumstances psychological treatment has physiological implications. Psychological treatments have had positive results for the treatment of NP in a few, small studies; however, treatment recommendations for NP have moved toward a multimodal approach incorporating psychological treatment with pharmacological or nonpharmacological methods (Turk et al., 2010). This serves as a possible opportunity that has not yet been thoroughly explored in the realm of PAP, through the combination of psychological and nonpharmacological treatment.

No single treatment method seems to be a superior method for alleviating PLP. This may be due to the nature of nonmechanism-based therapy development, treating symptoms rather than the root cause. In order to develop successful therapies, we should first seek to understand the primary mechanisms driving PLP in the background (Hsu and Cohen, 2013). We should also seek to understand the effects of various methods by reporting results in a consistent way. Several studies and the measured effects have been reported and reviewed; the unfortunate reality is that many of the therapy methods are difficult to compare in terms of effect because there is not a standard metric for PLP.

References

Agnew, W. F., McCreery, D. B., Yuen, T. G. and Bullara, L. A. (1999). Evolution and resolution of stimulation-induced axonal injury in peripheral nerve, *Muscle Nerve*, vol. 22, no. 10, pp. 1393–1402.

Alviar, M. J. M., Hale, T. and Dungca, M. (2016). Pharmacologic interventions for treating phantom limb pain, *Cochrane Database Syst. Rev.*, vol. 10, p. CD006380.

Barberá, J. and Albert-Pampló, R. (1993). Centrocentral anastomosis of the proximal nerve stump in the treatment of painful amputation neuromas of major nerves, *J. Neurosurg.*, vol. 79, no. 3, pp. 331–334.

Bekrater-Bodmann, R., Schredl, M., Diers, M., Reinhard, I., Foell, J., Trojan, J., Fuchs, X. and Flor, H. (2015). Post-amputation pain is associated with the recall of an impaired body representation in dreams – Results from a nation-wide survey on limb amputees, *PLoS ONE*, vol. 10, no. 3. doi: 10.1371/journal.pone.0119552.

Ben Abraham, R., Marouani, N. and Weinbroum, A. A. (2003). Dextromethorphan mitigates phantom pain in cancer amputees, *Ann. Surg. Oncol.*, vol. 10, no. 3, pp. 268–274.

Bennett, M. (2001). The LANSS Pain Scale: The Leeds assessment of neuropathic symptoms and signs, *Pain*, vol. 92, no. 1–2, pp. 147–157.

Blågestad, T., Pallesen, S., Grønli, J., Tang, N. K. Y. and Nordhus, I. H. (2016). How perceived pain influence sleep and mood more than the reverse: A novel, exploratory study with patients awaiting total hip arthroplasty, *Front. Psychol.*, vol. 7, p. 1689.

Blazer, D. G., Kessler, R. C., McGonagle, K. A. and Swartz, M. S. (1994). The prevalence and distribution of major depression in a national community sample: The National Comorbidity Survey, *Am. J. Psychiatry*, vol. 151, no. 7, pp. 979–986.

Bleakman, D., Alt, A. and Nisenbaum, E. S. (2006). Glutamate receptors and pain, *Semin. Cell Dev. Biol.*, vol. 17, no. 5, pp. 592–604.

Bone, M., Critchley, P. and Buggy, D. J. (2002). Gabapentin in postamputation phantom limb pain: A randomized, double-blind, placebo-controlled, cross-over study, *Reg. Anesth. Pain Med.*, vol. 27, no. 5, pp. 481–486.

Bouhassira, D., Attal, N., Fermanian, J., Alchaar, H., Gautron, M., Masquelier, E., Rostaing, S., Lanteri-Minet, M., Collin, E., Grisart, J. and Boureau, F. (2004). Development and validation of the neuropathic pain symptom inventory, *Pain*, vol. 108, no. 3, pp. 248–257.

Bouhassira, D., Attal, N., Alchaar, H., Boureau, F., Brochet, B., Bruxelle, J., Cunin, G., Fermanian, J., Ginies, P., Grun-Overdyking, A., Jafari-Schluep, H., Lantéri-Minet, M., Laurent, B., Mick, G., Serrie, A., Valade, D. and Vicaut, E. (2005). Comparison of pain syndromes associated with nervous or somatic lesions and development of a new neuropathic pain diagnostic questionnaire (DN4), *Pain*, vol. 114, no. 1–2, pp. 29–36.

Boutin, R. D., Pathria, M. N. and Resnick, D. (1998). Disorders in the stumps of amputee patients: MR imaging. *Am. J. Roentgenol.*, vol. 171, no. 2, pp. 497–501.

Bowering, K. J., O'Connell, N. E., Tabor, A., Catley, M. J., Leake, H. B., Moseley, G. L. and Stanton, T. R. (2013). The effects of graded motor imagery and its components on chronic pain: A systematic review and meta-analysis, *J. Pain Off. J. Am. Pain Soc.*, vol. 14, no. 1, pp. 3–13.

Brunelli, S., Morone, G., Iosa, M., Ciotti, C., De Giorgi, R., Foti, C. and Traballesi, M. (2015). Efficacy of progressive muscle relaxation, mental imagery, and phantom exercise training on phantom limb: A randomized controlled trial, *Arch. Phys. Med. Rehabil.*, vol. 96, no. 2, pp. 181–187.

Buonocore, M. (2015). Where is hidden the ghost in phantom sensations?, *World J. Clin. Cases*, vol. 3, no. 7, pp. 542–544.

Cansever, A., Uzun, O., Yildiz, C., Ates, A. and Atesalp, A. S. (2003). Depression in men with traumatic lower part amputation: A comparison to men with surgical lower part amputation, *Mil. Med.*, vol. 168, no. 2, pp. 106–109.

Casale, R., Alaa, L., Mallick, M. and Ring, H. (2009). Phantom limb related phenomena and their rehabilitation after lower limb amputation, *Eur. J. Phys. Rehabil. Med.*, vol. 45, no. 4, pp. 559–566.

Churchill, J. D., Arnold, L. L. and Garraghty, P. E. (2001). Somatotopic reorganization in the brainstem and thalamus following peripheral nerve injury in adult primates, *Brain Res.*, vol. 910, no. 1–2, pp. 142–152.

Cogan, S. F., Ludwig, K. A., Welle, C. G. and Takmakov, P. (2016). Tissue damage thresholds during therapeutic electrical stimulation, *J. Neural Eng.*, vol. 13, no. 2, p. 21001.

Collins, S., Sigtermans, M. J., Dahan, A., Zuurmond, W. W. A. and Perez, R. S. G. M. (2010). NMDA receptor antagonists for the treatment of neuropathic pain, *Pain Med.*, vol. 11, no. 11, pp. 1726–1742.

Dar, R., Ariely, D. and Frenk, H. (1995). The effect of past-injury on pain threshold and tolerance, *Pain*, vol. 60, no. 2, pp. 189–193.

Darnall, B. D., Ephraim, P., Wegener, S. T., Dillingham, T., Pezzin, L., Rossbach, P. and MacKenzie, E. J. (2005). Depressive symptoms and mental health service utilization among persons with limb loss: Results of a national survey, *Arch. Phys. Med. Rehabil.*, vol. 86, no. 4, pp. 650–658.

Davis E. N. and Chung, K. C. (2004). The Tinel sign: A historical perspective, *Plast. Reconstr. Surg.*, vol. 114, no. 2, pp. 494–499.

Devor M. and Wall, P. D. (1981). Plasticity in the spinal cord sensory map following peripheral nerve injury in rats, *J. Neurosci. Off. J. Soc. Neurosci.*, vol. 1, no. 7, pp. 679–684.

Dhillon, G. S., Lawrence, S. M., Hutchinson, D. T. and Horch, K. W. (2004). Residual function in peripheral nerve stumps of amputees: implications for neural control of artificial limbs, *J. Hand Surg.*, vol. 29, no. 4, pp. 605–615 (article), 616–618 (discussion).

Dillingham, T. R., Pezzin, L. E. and Shore, A. D. (2005). Reamputation, mortality, and health care costs among persons with dysvascular lower-limb amputations, *Arch. Phys. Med. Rehabil.*, vol. 86, no. 3, pp. 480–486.

Ducic, I., Mesbahi, A. N., Attinger, C. E. and Graw, K. (2008). The role of peripheral nerve surgery in the treatment of chronic pain associated with amputation stumps, *Plast. Reconstr. Surg.*, vol. 121, no. 3, pp. 908–914 (article), 915–917 (discussion).

Dworkin, R. H., Turk, D. C., Wyrwich, K. W., Beaton, D., Cleeland, C. S., Farrar, J. T., Haythornthwaite, J. A., Jensen, M. P., Kerns, R. D., Ader,

D. N., Brandenburg, N., Burke, L. B., Cella, D., Chandler, J., Cowan, P., Dimitrova, R., Dionne, R., Hertz, S., Jadad, A. R., Katz, N. P., Kehlet, H., Kramer, L. D., Manning, D. C., McCormick, C., McDermott, M. P., McQuay, H. J., Patel, S., Porter, L., Quessy, S., Rappaport, B. A., Rauschkolb, C., Revicki, D. A., Rothman, M., Schmader, K. E., Stacey, B. R., Stauffer, J. W., von Stein, T., White, R. E., Witter, J. and Zavisic, S. (2008). Interpreting the clinical importance of treatment outcomes in chronic pain clinical trials: IMMPACT recommendations, *J. Pain Off. J. Am. Pain Soc.*, vol. 9, no. 2, pp. 105–121.

Eccleston, C., Hearn, L. and Williams, A. C. de C. (2015). Psychological therapies for the management of chronic neuropathic pain in adults, *Cochrane Database Syst. Rev.*, no. 10, p. CD011259.

Ehde, D. M., Czerniecki, J. M., Smith, D. G., Campbell, K. M., Edwards, W. T., Jensen, M. P. and Robinson, L. R. (2000). Chronic phantom sensations, phantom pain, residual limb pain, and other regional pain after lower limb amputation, *Arch. Phys. Med. Rehabil.*, vol. 81, no. 8, pp. 1039–1044.

Eichenberger, U., Neff, F., Sveticic, G., Björgo, S., Petersen-Felix, S., Arendt-Nielsen, L. and Curatolo, M. (2008). Chronic phantom limb pain: the effects of calcitonin, ketamine, and their combination on pain and sensory thresholds, *Anesth. Analg.*, vol. 106, no. 4, pp. 1265–1273, table of contents.

Eklund, A., Nichols, T. E. and Knutsson, H. (2016). Cluster failure: Why fMRI inferences for spatial extent have inflated false-positive rates, *Proc. Natl. Acad. Sci. U. S. A.*, vol. 113, no. 28, pp. 7900–7905.

Ephraim, P. L., Wegener, S. T., MacKenzie, E. J., Dillingham, T. R. and Pezzin, L. E. (2005). Phantom pain, residual limb pain, and back pain in amputees: Results of a national survey, *Arch. Phys. Med. Rehabil.*, vol. 86, no. 10, pp. 1910–1919.

Finger, S. and Hustwit, M. P. (2003). Five early accounts of phantom limb in context: Paré, Descartes, Lemos, Bell, and Mitchell, *Neurosurgery*, vol. 52, no. 3, pp. 675–686.

Flor, H. (2002). Phantom-limb pain: Characteristics, causes, and treatment, *Lancet Neurol.*, vol. 1, no. 3, pp. 182–189.

Flor, H., Denke, C., Schaefer, M. and Grüsser, S. (2001). Effect of sensory discrimination training on cortical reorganisation and phantom limb pain, *The Lancet*, vol. 357, no. 9270, pp. 1763–1764.

Flor, H., Elbert, T., Knecht, S., Wienbruch, C., Pantev, C., Birbaumer, N., Larbig, W. and Taub, E. (1995). Phantom-limb pain as a perceptual correlate of cortical reorganization following arm amputation, *Nature*, vol. 375, no. 6531, pp. 482–484.

Flor, H., Mühlnickel, W., Karl, A., Denke, C., Grüsser, S., Kurth, R. and Taub, E. (2000). A neural substrate for nonpainful phantom limb phenomena, *Neuroreport*, vol. 11, no. 7, pp. 1407–1411.

Flor, H., Nikolajsen, L. and Staehelin Jensen, T. (2006). Phantom limb pain: A case of maladaptive CNS plasticity?, *Nat. Rev. Neurosci.*, vol. 7, no. 11, pp. 873–881.

Florence, S. L. and Kaas, J. H. (1995). Large-scale reorganization at multiple levels of the somatosensory pathway follows therapeutic amputation of the hand in monkeys, *J. Neurosci. Off. J. Soc. Neurosci.*, vol. 15, no. 12, pp. 8083–8095.

Forst, J. C., Blok, D. C., Slopsema, J. P., Boss, J. M., Heyboer, L. A., Tobias, C. M. and Polasek, K. H. (2015). Surface electrical stimulation to evoke referred sensation, *J. Rehabil. Res. Dev.*, vol. 52, no. 4, pp. 397–406.

Fried, K., Govrin-Lippmann, R., Rosenthal, F., Ellisman, M. H. and Devor, M. (1991). Ultrastructure of afferent axon endings in a neuroma, *J.Neurocytol.*, vol. 20, no. 8, pp. 682–701.

Galer, B. S. and Jensen, M. P. (1997). Development and preliminary validation of a pain measure specific to neuropathic pain: The Neuropathic Pain Scale, *Neurology*, vol. 48, no. 2, pp. 332–338.

Gallagher, P., Desmond, D. and MacLachlan, M. (Eds.,) (2008). *Psychoprosthetics*. London: Springer.

Garry, M. I., Loftus, A. and Summers, J. J. (2005). Mirror, mirror on the wall: viewing a mirror reflection of unilateral hand movements facilitates ipsilateral M1 excitability, *Exp. Brain Res.*, vol. 163, no. 1, pp. 118–122.

Gilron, I., Watson, C. P. N., Cahill, C. M. and Moulin, D. E. (2006). Neuropathic pain: A practical guide for the clinician, *CMAJ Can. Med. Assoc. J.*, vol. 175, no. 3, pp. 265–275.

Giummarra, M. J., Georgiou-Karistianis, N., Nicholls, M. E. R., Gibson, S. J., Chou, M. and Bradshaw, J. L. (2011). The menacing phantom: What pulls the trigger?, *Eur. J. Pain Lond. Engl.*, vol. 15, no. 7, p. 691. e1–8.

Giummarra, M. J., Gibson, S. J., Georgiou-Karistianis, N. and Bradshaw, J. L. (2007). Central mechanisms in phantom limb perception: The past, present and future, *Brain Res. Rev.*, vol. 54, no. 1, pp. 219–232.

Grüsser, S. M., Mühlnickel, W., Schaefer, M., Villringer, K., Christmann, C., Koeppe, C. and Flor, H. (2004). Remote activation of referred phantom sensation and cortical reorganization in human upper extremity amputees, *Exp. Brain Res.*, vol. 154, no. 1, pp. 97–102.

Halbert, J., Crotty, M. and Cameron, I. D. (2002). Evidence for the optimal management of acute and chronic phantom pain: A systematic review, *Clin. J. Pain*, vol. 18, no. 2, pp. 84–92.

Harris, A. J. (1999). Cortical origin of pathological pain, *Lancet Lond. Engl.*, vol. 354, no. 9188, pp. 1464–1466.

Hill, A. (1999). Phantom Limb Pain: A Review of the Literature on Attributes and Potential Mechanisms, *J. Pain Symptom Manage.*, vol. 17, no. 2, pp. 125–142.

House, J. W. and Brackmann, D. E. (1981). Tinnitus: Surgical treatment, *Ciba Found. Symp.*, vol. 85, pp. 204–216.

Hsu, E. and Cohen, S. P. (2013). Postamputation pain: epidemiology, mechanisms, and treatment, *J. Pain Res.*, vol. 6, pp. 121–136.

Huskisson, E. C. (1974). Measurement of Pain, *The Lancet*, vol. 304, no. 7889, pp. 1127–1131.

Iacono, R. P., Linford, J. and Sandyk, R. (1987). Pain management after lower extremity amputation, *Neurosurgery*, vol. 20, no. 3, pp. 496–500.

Jensen, M. P. and Karoly, P. Self-report scales and procedures for assessing pain in adults, in *Handbook of Pain Assessment*, Third., D. C. Turk and R. Melzack, Eds., New York, US: The Guilford Press, 2010, pp. 19–44.

Jensen, M. P., Smith, D. G., Ehde, D. M. and Robinsin, L. R. (2001). Pain site and the effects of amputation pain: Further clarification of the meaning of mild, moderate, and severe pain, *Pain*, vol. 91, no. 3, pp. 317–322.

Jensen, T. S., Krebs, B., Nielsen, J. and Rasmussen, P. (1983). Phantom limb, phantom pain and stump pain in amputees during the first 6 months following limb amputation, *Pain*, vol. 17, no. 3, pp. 243–256.

Jensen, T. S., Krebs, B., Nielsen, J. and Rasmussen, P. (1985). Immediate and long-term phantom limb pain in amputees: Incidence, clinical characteristics and relationship to pre-amputation limb pain, *Pain*, vol. 21, no. 3, pp. 267–278.

Johnson, C. (2016). Research Tools for the Measurement of Pain and Nociception, *Anim. Open Access J. MDPI*, vol. 6, no. 11.

Johnson, M. I., Mulvey, M. R. and Bagnall, A.-M. (2015). Transcutaneous electrical nerve stimulation (TENS) for phantom pain and stump pain following amputation in adults, *Cochrane Database Syst. Rev.*, vol. 8, p. CD007264.

Kashani, J. H., Frank, R. G., Kashani, S. R., Wonderlich, S. A., and Reid, J. C. (1983). Depression among amputees, *J. Clin. Psychiatry*, vol. 44, no. 7, pp. 256–258.

Katz, J. and Melzack, R. (1990). Pain 'memories' in phantom limbs: Review and clinical observations, *Pain*, vol. 43, no. 3, pp. 319–336.

Katz, J. and Melzack, R. (2010). The McGill pain questionnaire: Development psychometric properties, and usefulnes of the long form, short form, and short form-2, in *Handbook of Pain Assessment*, Third., D. C. Turk and R. Melzack, Eds., New York, US: The Guilford Press, pp. 45–66.

Katz, J. and Melzack, R. (1991). Auricular transcutaneous electrical nerve stimulation (TENS) reduces phantom limb pain, *J. Pain Symptom Manage.*, vol. 6, no. 2, pp. 73–83.

Katz, J. (1992). Psychophysiological contributions to phantom limbs, *Can. J. Psychiatry Rev. Can. Psychiatr.*, vol. 37, no. 5, pp. 282–298.

Kern, U., Busch, V., Müller, R., Kohl, M. and Birklein, F. (2012) Phantom limb pain in daily practice – Still a lot of work to do! *Pain Med. Malden Mass*, vol. 13, no. 12, pp. 1611–1626.

Kline, D. G. (2016). Silas Weir Mitchell and 'The Strange Case of George Dedlow,' *Neurosurg. Focus*, vol. 41, no. 1, p. E5.

Knežević, A., Salamon, T., Milankov, M., Ninković, S., Jeremić Knežević, M. and Tomašević Todorović, S. (2015). Assessment of quality of life in patients after lower limb amputation, *Med. Pregl.*, vol. 68, no. 3–4, pp. 103–108.

Kon, M. and Bloem, J. J. (1987). The treatment of amputation neuromas in fingers with a centrocentral nerve union, *Ann. Plast. Surg.*, vol. 18, no. 6, pp. 506–510.

Kooijman, C. M., Dijkstra, P. U., Geertzen, J. H., Elzinga, A. and van der Schans, C. P. (2000). Phantom pain and phantom sensations in upper limb amputees: An epidemiological study, *Pain*, vol. 87, no. 1, pp. 33–41.

Krause, S. J. and Backonja, M.-M. (2003). Development of a neuropathic pain questionnaire, *Clin. J. Pain*, vol. 19, no. 5, pp. 306–314.

Lacey, H. P., Fagerlin, A., Loewenstein, G., Smith, D. M., Riis, J. and Ubel, P. A. (2008). Are they really that happy? Exploring scale recalibration in estimates of well-being, *Health Psychol. Off. J. Div. Health Psychol. Am. Psychol. Assoc.*, vol. 27, no. 6, pp. 669–675.

Legro, M. W., Reiber, G. D., Smith, D. G., del Aguila, M., Larsen, J. and Boone, D. (1998). Prosthesis evaluation questionnaire for persons with lower limb amputations: Assessing prosthesis-related quality of life, *Arch. Phys. Med. Rehabil.*, vol. 79, no. 8, pp. 931–938.

Limakatso, K., Corten, L. and Parker, R. (2016). The effects of graded motor imagery and its components on phantom limb pain and disability in upper and lower limb amputees: A systematic review protocol, *Syst. Rev.*, vol. 5, p. 145.

Liu, T. T. and Frank, L. R. (2004). Efficiency, power and entropy in event-related FMRI with multiple trial types. Part I: theory, *NeuroImage*, vol. 21, no. 1, pp. 387–400.

Liu, T. T. (2004). Efficiency, power, and entropy in event-related fMRI with multiple trial types. Part II: design of experiments, *NeuroImage*, vol. 21, no. 1, pp. 401–413.

Lotze, M., Flor, H., Grodd, W., Larbig, W. and Birbaumer, N. (2001). Phantom movements and pain. An fMRI study in upper limb amputees, *Brain J. Neurol.*, vol. 124, no. Pt 11, pp. 2268–2277.

Lu, M.-C., Ho, C.-Y., Hsu, S.-F., Lee, H.-C., Lin, J.-H., Yao, C.-H. and Chen, Y.-S. (2008). Effects of electrical stimulation at different frequencies on regeneration of transected peripheral nerve, *Neurorehabil. Neural Repair*, vol. 22, no. 4, pp. 367–373.

Mackey, S., Carroll, I., Emir, B., Murphy, T. K., Whalen, E. and Dumenci, L. (2012). Sensory pain qualities in neuropathic pain, *J. Pain Off. J. Am. Pain Soc.*, vol. 13, no. 1, pp. 58–63.

Mackinnon, S. E. and Dellon, A. L. (1987). Results of treatment of recurrent dorsoradial wrist neuromas, *Ann. Plast. Surg.*, vol. 19, no. 1, pp. 54–61.

Mackinnon, S. E., Dellon, A. L., Hudson, A. R. and Hunter, D. A. (1985). Alteration of neuroma formation by manipulation of its microenvironment, *Plast. Reconstr. Surg.*, vol. 76, no. 3, pp. 345–353.

Macuga, K. L. and Frey, S. H. (2012). Neural representations involved in observed, imagined, and imitated actions are dissociable and hierarchically organized, *NeuroImage*, vol. 59, no. 3, pp. 2798–2807.

Marshall, H. M., Jensen, M. P., Ehde, D. M. and Campbell, K. M. (2002). Pain site and impairment in individuals with amputation pain, *Arch. Phys. Med. Rehabil.*, vol. 83, no. 8, pp. 1116–1119.

Matsubayashi, Y., Takeshita, K., Sumitani, M., Oshima, Y., Tonosu, J., Kato, S., Ohya, J., Oichi, T., Okamoto, N. and Tanaka, S. (2015). Psychometric validation of the Japanese version of the neuropathic pain symptom inventory, *PLoS ONE*, vol. 10, no. 11: e0143350. https://doi.org/10.1371/journal.pone.0143350.

McCabe, C. S., Haigh, R. C., Halligan, P. W. and Blake, D. R. (2003). Referred sensations in patients with complex regional pain syndrome type 1, *Rheumatol. Oxf. Engl.*, vol. 42, no. 9, pp. 1067–1073.

McCreery, D. B., Agnew, W. F., Yuen, T. G. and Bullara, L. A. (1995). Relationship between stimulus amplitude, stimulus frequency and neural damage during electrical stimulation of sciatic nerve of cat, *Med. Biol. Eng. Comput.*, vol. 33, no. 3 Spec No, pp. 426–429.

McCreery, D. B., Agnew, W. F., Yuen, T. G. and Bullara, L. A. (1992). Damage in peripheral nerve from continuous electrical stimulation: Comparison of two stimulus waveforms, *Med. Biol. Eng. Comput.*, vol. 30, no. 1, pp. 109–114.

McCreery, D. B., Yuen, T. G., Agnew, W. F. and Bullara, L. A. (1997). A characterization of the effects on neuronal excitability due to prolonged microstimulation with chronically implanted microelectrodes, *IEEE Trans. Biomed. Eng.*, vol. 44, no. 10, pp. 931–939.

Melzack, R. and Wall, P. D. (1965). Pain mechanisms: A new theory, *Science*, vol. 150, no. 3699, pp. 971–979.

Melzack, R. (1999). From the gate to the neuromatrix, *Pain*, vol. Suppl 6, pp. S121–126.

Melzack, R. (1992). Phantom limbs, *Sci. Am.*, vol. 266, no. 4, pp. 120–126.

Melzack, R. (1975). The McGill pain questionnaire: Major properties and scoring methods, *PAIN*, vol. 1, no. 3, pp. 277–299.

Melzack, R. (1990). Phantom limbs and the concept of a neuromatrix, *Trends Neurosci.*, vol. 13, no. 3, pp. 88–92.

Melzack, R., Israel, R., Lacroix, R. and Schultz, G. (1997). Phantom limbs in people with congenital limb deficiency or amputation in early childhood, *Brain J. Neurol.*, vol. 120 (Pt 9), pp. 1603–1620.

Mendell, L. M. (2014). Constructing and deconstructing the Gate theory of pain, *Pain*, vol. 155, no. 2, pp. 210–216.

Mercier, C. and Sirigu, A. (2009). Training with virtual visual feedback to alleviate phantom limb pain, *Neurorehabil. Neural Repair*, vol. 23, no. 6, pp. 587–594.

Montoya, P., Larbig, W., Grulke, N., Flor, H., Taub, E. and Birbaumer, N. (1997). The relationship of phantom limb pain to other phantom limb phenomena in upper extremity amputees, *Pain*, vol. 72, no. 1–2, pp. 87–93.

Moore, C. I., Stern, C. E., Dunbar, C., Kostyk, S. K., Gehi, A. and Corkin, S., (2000). Referred phantom sensations and cortical reorganization after spinal cord injury in humans, *Proc. Natl. Acad. Sci. U. S. A.*, vol. 97, no. 26, pp. 14703–14708.

Mortimer, C. M., MacDonald, R. J. M., Martin, D. J., McMillan, I. R., Ravey, J. and Steedman, W. M. (2004). A focus group study of health professionals' views on phantom sensation, phantom pain and the need for patient information, *Patient Educ. Couns.*, vol. 54, no. 2, pp. 221–226.

Moseley, G. L. (2006). Graded motor imagery for pathologic pain: A randomized controlled trial, *Neurology*, vol. 67, no. 12, pp. 2129–2134.

Nikolajsen, L., Ilkjaer, S., Krøner, K., Christensen, J. H. and Jensen, T. S. (1997). The influence of preamputation pain on postamputation stump and phantom pain, *Pain*, vol. 72, no. 3, pp. 393–405.

Nikolajsen, L., Black, J. A., Kroner, K., Jensen, T. S. and Waxman, S. G. (2010). Neuroma removal for neuropathic pain: Efficacy and predictive value of lidocaine infusion, *Clin. J. Pain*, vol. 26, no. 9, pp. 788–793.

Nikolajsen, L., Hansen, C. L., Nielsen, J., Keller, J., Arendt-Nielsen, L. and Jensen, T. S. (1996). The effect of ketamine on phantom pain: A central neuropathic disorder maintained by peripheral input, *Pain*, vol. 67, no. 1, pp. 69–77.

Nystrom, B. and Hagbarth, K. E. (1981). Microelectrode recordings from transected nerves in amputees with phantom limb pain, *Neurosci.Lett.*, vol. 27, no. 2, pp. 211–216.

O'Reilly, M. A. R., O'Reilly, P. M. R., O'Reilly, H. M. R., Sullivan, J. and Sheahan, J. (2013). High-resolution ultrasound findings in the symptomatic residual limbs of amputees, *Mil. Med.*, vol. 178, no. 12, pp. 1291–1297.

O'Reilly, M. A. R., O'Reilly, P. M. R., Sheahan, J. N., Sullivan, J., O'Reilly, H. M. and O'Reilly, M. J. (2016). Neuromas as the cause of pain in the residual limbs of amputees. An ultrasound study, *Clin. Radiol.*, vol. 71, no. 10, p. 1068.e1–1068.e6.

Osborne, T. L., Raichle, K. A., Jensen, M. P., Ehde, D. M. and Kraft, G. (2006). The reliability and validity of pain interference measures in persons with multiple sclerosis, *J. Pain Symptom Manage.*, vol. 32, no. 3, pp. 217–229.

Owings, M. F. and Kozak, L. J. (1998). Ambulatory and inpatient procedures in the United States, 1996, *Vital Health Stat.*, vol. 13, no. 139, pp. 1–119.

Parry, C. B. (1980). Pain in avulsion lesions of the brachial plexus, *Pain*, vol. 9, no. 1, pp. 41–53.

Patel, Y. A. and Butera, R. J. (2015). Differential fiber-specific block of nerve conduction in mammalian peripheral nerves using kilohertz electrical stimulation, *J. Neurophysiol.*, vol. 113, no. 10, pp. 3923–3929.

Priganc, V. W. and Stralka, S. W. (2011). Graded motor imagery, *J. Hand Ther.*, vol. 24, no. 2, pp. 164–169.

Prodanov, D., Marani, E. and Holsheimer, J. (2003). Functional electric stimulation for sensory and motor functions: progress and problems, *Biomed. Rev.*, vol. 14, no. 0, pp. 23–50.

Qiu, T., Chen, L., Mao, Y., Wu, J., Tang, W., Hu, S., Zhou, L. and Gu, Y. (2014). Sensorimotor cortical changes assessed with resting-state fMRI following total brachial plexus root avulsion, *J. Neurol. Neurosurg. Psychiatry*, vol. 85, no. 1, pp. 99–105.

Raichle, K. A., Osborne, T. L., Jensen, M. P. and Cardenas, D. (2006). The reliability and validity of pain interference measures in persons with spinal cord injury, *J. Pain Off. J. Am. Pain Soc.*, vol. 7, no. 3, pp. 179–186.

Ramachandran, V. S. and Rogers-Ramachandran, D. (1996). Synaesthesia in phantom limbs induced with mirrors, *Proc. Biol. Sci.*, vol. 263, no. 1369, pp. 377–386.

Ramachandran, V. S., Rogers-Ramachandran, D. and Stewart, M. (1992). Perceptual correlates of massive cortical reorganization, *Science*, vol. 258, no. 5085, pp. 1159–1160.

Rauck, R. L., Cohen, S. P., Gilmore, C. A., North, J. M., Kapural, L., Zang, R. H., Grill J. H. and Boggs, J. W. (2014). Treatment of post-amputation pain with peripheral nerve stimulation, *Neuromodulation J. Int. Neuromodulation Soc.*, vol. 17, no. 2, pp. 188–197.

Reilly, K. T., Mercier, C., Schieber, M. H. and Sirigu, A. (2006). Persistent hand motor commands in the amputees' brain, *Brain J. Neurol.*, vol. 129, no. Pt 8, pp. 2211–2223.

Robinson, L. R., Czerniecki, J. M., Ehde, D. M., Edwards, W. T., Judish, D. A., Goldberg, M. L., Campbell, K. M., Smith, D. G. and Jensen, M. P. (2004). Trial of amitriptyline for relief of pain in amputees: results of a randomized controlled study, *Arch. Phys. Med. Rehabil.*, vol. 85, no. 1, pp. 1–6.

Roed Rasmussen, M. L., Prause, J. U., Johnson, M. and Toft, P. B. (2009). Phantom eye syndrome: Types of visual hallucinations and related phenomena, *Ophthal. Plast. Reconstr. Surg.*, vol. 25, no. 5, pp. 390–393.

Sörös, P., Knecht, S., Bantel, C., Imai, T., Wüsten, R., Pantev, C., Lütkenhöner, B., Bürkle, H. and Henningsen, H. (2001). Functional reorganization of the human primary somatosensory cortex after acute pain demonstrated by magnetoencephalography, *Neurosci. Lett.*, vol. 298, no. 3, pp. 195–198.

Schley, M. T., Wilms, P., Toepfner, S., Schaller, H.-P., Schmelz, M., Konrad, C. J. and Birbaumer, N. (2008). Painful and nonpainful phantom and stump sensations in acute traumatic amputees, *J. Trauma*, vol. 65, no. 4, pp. 858–864.

Schwenkreis, P., Pleger, B., Cornelius, B., Weyen, U., Dertwinkel, R., Zenz, M., Malin, J. P. and Tegenthoff, M. (2003). Reorganization in the ipsilateral motor cortex of patients with lower limb amputation, *Neurosci. Lett.*, vol. 349, no. 3, pp. 187–190.

Sehirlioglu, A., Ozturk, C., Yazicioglu, K., Tugcu, I., Yilmaz, B. and Goktepe, A. S. (2009). Painful neuroma requiring surgical excision after lower limb amputation caused by landmine explosions, *Int. Orthop.*, vol. 33, no. 2, pp. 533–536.

Sherman, R. A. and Sherman, C. J. (1983). Prevalence and characteristics of chronic phantom limb pain among American veterans. Results of a trial survey, *Am. J. Phys. Med.*, vol. 62, no. 5, pp. 227–238.

Sherman, R. A. (1980). Published treatments of phantom limb pain, *Am. J. Phys. Med.*, vol. 59, no. 5, pp. 232–244.

Sherman, R. A., Sherman, C. J. and Parker, L. (1984). Chronic phantom and stump pain among American veterans: Results of a survey, *Pain*, vol. 18, no. 1, pp. 83–95.

Sherman, R. A., Sherman, C. J. and Bruno, G. M. (1987). Psychological factors influencing chronic phantom limb pain: An analysis of the literature, *Pain*, vol. 28, no. 3, pp. 285–295.

Sherman, R. A., Arena, J. G., Sherman, C. J. and Ernst, J. L. (1989). The mystery of phantom pain: Growing evidence for psychophysiological mechanisms, *Biofeedback Self-Regul.*, vol. 14, no. 4, pp. 267–280.

Sherman, R. A., Sherman, C. J. and Gall, N. G. (1980). A survey of current phantom limb pain treatment in the United States, *Pain*, vol. 8, no. 1, pp. 85–99.

Shore, S. E., Roberts, L. E. and Langguth, B. (2016). Maladaptive plasticity in tinnitus–Triggers, mechanisms and treatment, *Nat. Rev. Neurol.*, vol. 12, no. 3, pp. 150–160.

Shukla, G. D., Sahu, S. C., Tripathi, R. P. and Gupta, D. K. (1982). A psychiatric study of amputees, *Br. J. Psychiatry J. Ment. Sci.*, vol. 141, pp. 50–53.

Smith, D. G., Ehde, D. M., Hanley, M. A., Campbell, K. M., Jensen, M. P., Hoffman, A. J., Awan, A. B., Czerniecki, J. M. and Robinson, L. R. (2005). Efficacy of gabapentin in treating chronic phantom limb and residual limb pain, *J. Rehabil. Res. Dev.*, vol. 42, no. 5, pp. 645–654.

Smith, D. G., Ehde, D. M., Legro, M. W., Reiber, G. E., del Aguila, M. and Boone, D. A. (1999). Phantom limb, residual limb, and back pain after lower extremity amputations, *Clin. Orthop.*, no. 361, pp. 29–38.

Soler, M. D., Kumru, H., Vidal, J., Pelayo, R., Tormos, J. M., Fregni, F., Navarro, X. and Pascual-Leone, A. (2010). Referred sensations and neuropathic pain following spinal cord injury, *Pain*, vol. 150, no. 1, pp. 192–198.

Sommer, C., Richter, H., Rogausch, J. P., Frettlöh, J., Lungenhausen, M. and Maier, C. (2011). A modified score to identify and discriminate neuropathic pain: A study on the German version of the neuropathic pain symptom inventory (NPSI), *BMC Neurol.*, vol. 11, p. 104.

Son, B.-C. and Ha, S.-W. (2015). Phantom remodeling effect of dorsal root entry zone lesioning in phantom limb pain caused by brachial plexus avulsion, *Stereotact. Funct. Neurosurg.*, vol. 93, no. 4, pp. 240–244.

Souza, J. M., Cheesborough, J. E., Ko, J. H., Cho, M. S., Kuiken, T. A. and Dumanian, G. A. (2014). Targeted muscle reinnervation: A novel approach to postamputation neuroma pain, *Clin. Orthop.*, vol. 472, no. 10, pp. 2984–2990.

Sun, Q., Tu, H., Xing, G.-G., Han, J.-S. and Wan, Y. (2005). Ectopic discharges from injured nerve fibers are highly correlated with tactile allodynia only in early, but not late, stage in rats with spinal nerve ligation, *Exp. Neurol.*, vol. 191, no. 1, pp. 128–136.

Swanson, A. B., Boeve, N. R. and Lumsden, R. M. (1977). The prevention and treatment of amputation neuromata by silicone capping, *J. Hand Surg.*, vol. 2, no. 1, pp. 70–78.

Sweet, W. H. (1975). 'Phantom' sensations following intraspinal injury, *Neurochirurgia (Stuttg.)*, vol. 18, no. 5, pp. 139–154.

Tan, G., Jensen, M. P., Thornby, J. I. and Shanti, B. F. (2004). Validation of the brief pain inventory for chronic nonmalignant pain, *J. Pain Off. J. Am. Pain Soc.*, vol. 5, no. 2, pp. 133–137.

Teixeira, M. J., da, M. G., da Paz, S., Bina, M. T., Santos, S. N., Raicher, I., Galhardoni, R., Fernandes, D. T., Yeng, L. T., Baptista, A. F. and de Andrade, D. C. (2015). Neuropathic pain after brachial plexus avulsion–Central and peripheral mechanisms, *BMC Neurol.*, vol. 15. no. 73. doi: 10.1186/s12883-015-0329-x.

Thompson, A. K. and Wolpaw, J. R. (2014). Operant conditioning of spinal reflexes: From basic science to clinical therapy, *Front. Integr. Neurosci.*, vol. 8. no. 25. doi: 10.3389/fnint.2014.00025

Tilak, M., Isaac, S. A., Fletcher, J., Vasanthan, L. T., Subbaiah, R. S., Babu, A., Bhide, R. and Tharion, G. (2016). Mirror therapy and transcutaneous electrical nerve stimulation for management of phantom limb pain in amputees – A single blinded randomized controlled trial, *Physiother. Res. Int. J. Res. Clin. Phys. Ther.*, vol. 21, no. 2, pp. 109–115.

Trevelyan, E. G., Turner, W. A. and Robinson, N. (2016). Perceptions of phantom limb pain in lower limb amputees and its effect on quality of life: A qualitative study, *Br. J. Pain*, vol. 10, no. 2, pp. 70–77.

Trotter, W. and Davies, H. M. (1909).Experimental studies in the innervation of the skin, *J. Physiol.*, vol. 38, no. 2–3, p. 134–246.1.

Tsao, J. W., Finn, S. B. and Miller, M. E. (2016). Reversal of phantom pain and hand-to-face remapping after brachial plexus avulsion, *Ann. Clin. Transl. Neurol.*, vol. 3, no. 6, pp. 463–464.

Tupper, J. W. and Booth, D. M. (1976). Treatment of painful neuromas of sensory nerves in the hand: A comparison of traditional and newer methods, *J. Hand Surg.*, vol. 1, no. 2, pp. 144–151.

Turk, D. C., Audette, J., Levy, R. M., Mackey, S. C. and Stanos, S. (2010). Assessment and treatment of psychosocial comorbidities in patients with neuropathic pain, *Mayo Clin. Proc.*, vol. 85, no. 3 Suppl, pp. S42–S50.

Vernadakis, A. J., Koch, H. and Mackinnon, S. E. (2003). Management of neuromas, *Clin. Plast. Surg.*, vol. 30, no. 2, pp. 247–268.

Von Korff, M., Ormel, J., Keefe, F. J. and Dworkin, S. F. (1992). Grading the severity of chronic pain, *Pain*, vol. 50, no. 2, pp. 133–149.

Wall, P. D. (1978). The gate control theory of pain mechanisms. A re-examination and re-statement, *Brain J. Neurol.*, vol. 101, no. 1, pp. 1–18.

Wartan, S. W., Hamann, W., Wedley, J. R. and McColl, I. (1997). Phantom pain and sensation among British veteran amputees, *Br. J. Anaesth.*, vol. 78, no. 6, pp. 652–659.

Wasner, G. L. and Brock, J. A. (2008). Determinants of thermal pain thresholds in normal subjects, *Clin. Neurophysiol. Off. J. Int. Fed. Clin. Neurophysiol.*, vol. 119, no. 10, pp. 2389–2395.

Wu, C. W. H. and Kaas, J. H. (2002). The effects of long-standing limb loss on anatomical reorganization of the somatosensory afferents in the brainstem and spinal cord, *Somatosens. Mot. Res.*, vol. 19, no. 2, pp. 153–163.

Wu, C. L., Agarwal, S., Tella, P. K., Klick, B., Clark, M. R., Haythornthwaite, J. A., Max, M. B. and Raja, S. N. (2008). Morphine versus mexiletine for treatment of postamputation pain, *Anesthesiology*, vol. 109, no. 2, pp. 289–296.

Wu, C. L., Tella, P., Staats, P. S., Vaslav, R., Kazim, D. A., Wesselmann, U. and Raja, S. N. (2002). Analgesic effects of intravenous lidocaine and morphine on postamputation pain: A randomized double-blind, active placebo-controlled, crossover trial, *Anesthesiology*, vol. 96, no. 4, pp. 841–848.

Xu, J. and Wall, J. T. (1997). Rapid changes in brainstem maps of adult primates after peripheral injury, *Brain Res.*, vol. 774, no. 1–2, pp. 211–215.

Yanagisawa, T., Fukuma, R., Seymour, B., Hosomi, K., Kishima, H., Shimizu, T., Yokoi, H., Hirata, M., Yoshimine, T., Kamitani, Y. and Saitoh, Y. (2016). Induced sensorimotor brain plasticity controls pain in phantom limb patients, *Nat. Commun.*, vol. 7, p. 13209.

Zhao, J., Guo, X., Xia, X., Peng, W., Wang, W., Li, S., Zhang, Y. and Hu, L. (2016). Functional reorganization of the primary somatosensory cortex of a phantom limb pain patient, *Pain Physician*, vol. 19, no. 5, pp. E781–E786.

Ziegler-Graham, K., MacKenzie, E. J., Ephraim, P. L., Travison, T. G. and Brookmeyer, R. (2008). Estimating the prevalence of limb loss in the United States: 2005 to 2050, *Arch. Phys. Med. Rehabil.*, vol. 89, no. 3, pp. 422–429.

2

Neurobiology of Pain

Víctor M. López-Álvarez[1,2], Elena Redondo-Castro[1,2] and Xavier Navarro[1,2,*]

[1]Institute of Neurosciences, Department of Cell Biology, Physiology and Immunology, Universitat Autònoma de Barcelona, Bellaterra, Spain
[2]Centro de Investigación Biomédica en Red Sobre Enfermedades Neurodegenerativas (CIBERNED), Bellaterra, Spain
E-mail: xavier.navarro@uab.cat
*Corresponding Author

2.1 Physiology of Pain

Pain is a physiological experience, designed to alert us from potential damages to our body, so it has a clear protective role. Pain is defined by the IASP (International Association for the Study of Pain) as *an unpleasant sensory and emotional experience associated with actual or potential tissue damage, or described in terms of such damage*. When the pain circuits are correctly working they aware us from external (abnormal heating, pinch stimuli, etc.) or internal stimuli (cardiac ischemia) that would potentially hurt the tissues. Ideally, the sensation we perceive should be unpleasant enough, so it cannot be ignored, and the sensation should continue as long as the stimulus is present. Different types of "normal" pain can be distinguished depending on their origin and characteristics: acute (or pricking), chronic (or burning), and continuous or visceral.

2.1.1 Nociceptors and Nociceptive Fibers

There are two main types of nerve fibers conveying pain information: C fibers and Aδ fibers (Table 2.1). In both cases, the stimuli may come from the skin, muscle, and joint tissues or certain visceral structures. They do not present a clear ending receptor structure and are commonly identified as free nerve endings.

Table 2.1 Classification of the primary afferent axons in the peripheral nervous system

Fiber	Myelin	Diameter (μm)	Velocity (m/s)	Function
Aα	Yes	13–20	80–120	Proprioceptors of skeletal muscle
Aβ	Yes	6–12	35–75	Touch, mechanoreceptors, proprioceptors
Aδ	Yes	1–5	5–30	Pain, mechanical and thermal receptors
C	No	0.2–1.5	0.5–2	Pain, warm thermoreceptors

Aδ fibers are thinly myelinated, so they can conduct a *fast pain* signal, at 5–30 m/s. In this case, Aδ nociceptive fibers convey nociceptive information as well as information coming from intense mechanical or thermal stimulation. This fast pain has been reported as the *first pain*, the initial sharp painful sensation just after the contact with the noxious stimuli.

C fibers are related with a *slow pain*, since these unmyelinated fibers conduct impulses at less than 2 m/s, normally evoked by thermal, mechanical, and chemical stimuli. Most of them act as **polymodal nociceptors**, although a proportion seems to be sensitive only to mechanical or thermal stimuli. This slow pain is also called *second pain*, and evokes a more diffuse and longer lasting painful sensation than the pain evoked by the Aδ fibers.

The unmyelinated C nociceptive afferents can be divided into two major subpopulations, the peptidergic and the nonpeptidergic. The peptidergic nociceptors express substance P (SP) and calcitonin gene-related peptide (CGRP), and the nonpeptidergic ones possess fluoride-resistant acid phosphatase (FRAP) activity, bind the lectin IB4 and express purinergic P2X3 receptors. These two populations differ in neurotrophic support in the adult. In fact, during development, both populations require nerve growth factor (NGF) for survival, but shortly after birth only the peptidergic continue to respond to NGF, whereas the nonpeptidergic population starts to respond instead to glial cell line-derived neurotrophic factor (GDNF). Accordingly, the peptidergic population expresses the NGF high-affinity receptor trkA, whereas the nonpeptidergic expresses GDNF receptors. Although the distinction between two populations of primary sensory fibers, peptidergic and nonpeptidergic, seems attractive, it is not completely accurate as a small proportion of peptidergic sensory fibers (those that colocalize CGRP and somatostatin (SOM)) do not respond to NGF in the adult and bind the lectin IB4. In the spinal cord, sensory fibers that express SP and CGRP terminate mostly in laminae I, outer II, and V; those that colocalize CGRP and SOM terminate in laminae I and II, and those that contain FRAP, bind the lectin IB4 or express the purinergic P2X3 receptor terminate mostly in the inner lamina II.

The peripheral terminal of the nociceptor, embedded in the tissue, is where the noxious stimuli are detected and transduced into electrical impulses. This leads to the train of events that allows for the conscious sensation of pain. The sensory specificity of nociceptors is established by the high threshold to particular types of stimuli. Only when the high threshold has been reached by either chemical, thermal, or mechanical events are the nociceptors activated.

The transduction of the nociceptive information starts in the periphery, where a stimulus is able to activate the nociceptor endings, by stretching or bending the nociceptor surface or by activating ion channels present in its membrane. At the site of injury some algogenous substances are released, such as proteases, bradykinin, ATP, and potassium ions. Due to the variety of stimuli that can elicit nociceptive signaling (thermal, mechanical, and chemical stimuli), different specific receptors have been described. Both Aδ and C fibers present transient receptor potential (TRP) receptors, which resemble voltage gated ion channels, presenting six transmembrane domains and a central pore which allows an influx of sodium and calcium that initiates the generation of action potentials. One subtype of these channels is the vanilloid receptor (TRPV1), that responds to capsaicin but is also activated by acidification of the extracellular medium, by heat (thermal threshold $>43°C$) and by anandamide, an endogenous cannabinoid. Other channels of the TRP family activated by temperature increase are TRPV3 and TRPV4, with thresholds of 33°C and 25°C, respectively. There are also nociceptors activated by intense cold, below 15°C, that causes painful sensations with a burning or stabbing component. The TRPA1 channel is activated both by temperatures below 18°C and by irritant compounds.

Other channels present in nociceptive terminals are those of the degenerin family (Deg/ENaC), which are epithelial sodium channels activated by mechanical forces. Some of these channels also respond to decrease in the pH of the extracellular medium and are therefore called acid sensing channels (ASICs). Also important as stimulus transducer channels are those belonging to a diverse family of channels for potassium ions (K_{2P} channels). These channels are open at rest, and are closed by the stimulus, causing depolarization of the fiber. Their activity can be modified by mechanical, thermal stimuli, intracellular and extracellular pH changes, and hypoxia.

2.1.2 Nociceptive Spinal Cord Circuits

The different subpopulations of primary afferent fibers convey nociceptive information through parallel spinal pathways. At every stage of the pain pathway – from sensory nerve to spinal cord, from spinal cord to brainstem and from brainstem to the forebrain – information signaling injury is subdivided

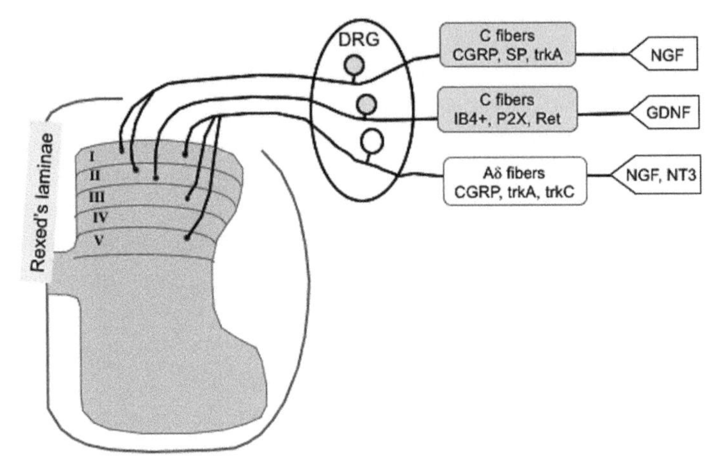

Figure 2.1 Representation of the three main types of nociceptive fibers, their markers and the laminae of the dorsal horn where they mainly project central terminals.

or shared between these parallel systems. Molecular dissection has begun to reveal distinct functions for these separate pathways and their contribution to the final behavioral outcome.

Nociceptive neurons have their body in the dorsal root ganglia, where they can be divided in three main populations (Figure 2.1): (i) mid-size neurons with Aδ axons, that express CGRP, and TRPV2 receptors, (ii) small neurons with unmyelinated C axons, expressing neuropeptides (SP, CGRP) and TRPV1 receptors, and (iii) small neurons nonpeptidergic, with C axons, that express receptors TRPV1 and P2X. It has been suggested that the C peptidergic pathway would be particularly important in conveying inflammatory pain, whereas the nonpeptidergic pathway would play a role in neuropathic pain.

Peptidergic primary afferents connect with the pathway that derives largely from lamina I neurons of the dorsal horn expressing the NK1 receptor, and terminates within the thalamus, the parabrachial area, and the periaqueductal grey matter (PAG). These latter areas in turn project on brain areas such as the hypothalamus and amygdala that modulate the affective dimensions of pain and control autonomic activity. The lamina I pathway is mainly involved in signaling the intensity of pain, therefore these second-order neurons are capable of reliably detecting and transmitting precise quantitative information about noxious pressure and noxious heat to higher centers. Of interest, the selective destruction of lamina I results in loss of the increased sensitivity to stimulation that follows inflammation or manipulation of the peripheral nerve.

On the other hand, the IB4 binding subpopulation of nonpeptidergic C fibers contact with neurons in inner lamina II of the spinal cord. These interneurons in turn contact projection neurons of lamina V, and many of these send axons to fourth-order neurons in the amygdala, hypothalamus, bed nucleus of the stria terminalis and globus pallidus. Interneurons in this lamina also show increase in protein kinase Cgamma (PKC_{γ}) following inflammation which results in mechanical hypersensitivity. However, these interneurons predominantly receive inputs from myelinated, rather than unmyelinated, primary afferent terminals. In contrast, calbindin positive interneurons of lamina IIi are located postsynaptic to the IB4-positive subpopulation of nonpeptidergic C fibers, but not to myelinated afferents.

Type A nociceptive fibers send collateral branches to contact with the wide dynamic range (WDR) neurons or nociceptive neurons type 2, located mainly in lamina V. Taken together, these observations illustrate the very complex connectivity of primary afferents with the interneurons of the dorsal horn of the spinal cord.

2.1.3 Nociceptive Ascending Pathways

The nociceptive information arriving from the periphery travels along the peripheral axonal branch of primary nociceptive neurons, whose soma are located in the dorsal root ganglia, and the central axon entering into the spinal cord by the dorsal roots. After the dorsal root entry, they travel within the zone of Lissauer, in which axons move up or down a pair of segments before entering the gray matter of the dorsal horn, in a region called substantia gelatinosa. Central nociceptive terminals contact to second-order neurons mainly placed in laminae I and II (pure nociceptive), and V (mixed nociceptive and mecanosensory, WDR). Sensory fibers that are peptidergic terminate mostly in laminae I and outer II, and a few in lamina V; those that are nonpeptidergic terminate mostly in the inner lamina II. The main neurotransmitter involved in these first relay is glutamate, but also substance P, acting as cotransmitter in peptidergic nociceptors, is important to experience moderate to intense pain.

From the second-order neuron the thermal and nociceptive information crosses the midline and ascends to the brain in the spinothalamic tract, part of the anterolateral system. This decussation occurs at the spinal level and in two or three segments all the fibers are in the contralateral side. The ascending axons travel through the medulla, the pons and the midbrain without synapsing, until reaching the thalamus. From here, the information is conveyed to the primary somatosensory cortex (Figure 2.2). This route

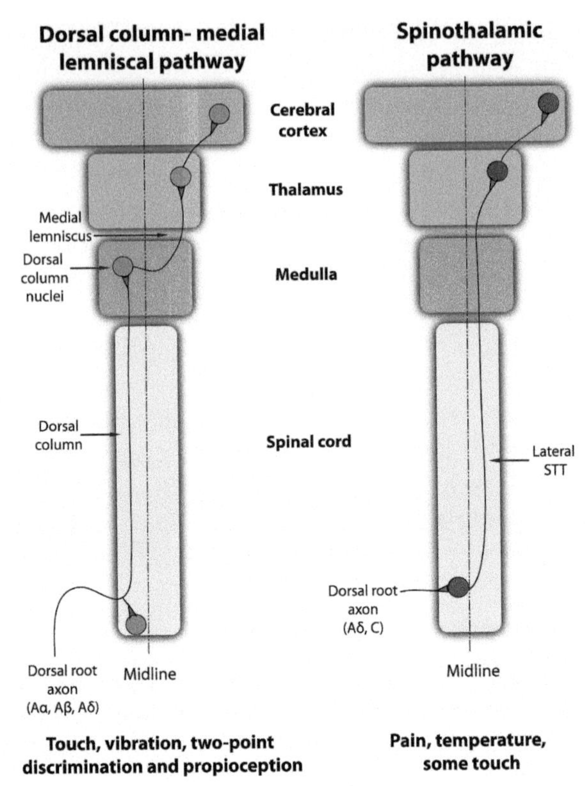

Figure 2.2 Schematic of the ascending pathways of somatic sensations. Low threshold mechanoreceptive afferents follow the dorsal column system, that decussates at the medulla, whereas thermal and pain afferents constitute the anterolateral system, crossed at segmental cord levels.

is followed in order to transmit the gross information of pain, the essential information for the brain to note stimuli that threaten the integrity of the body.

Axons from the second-order spinal neurons make relays on different structures in order to mediate different aspects of the sensory and behavioral response to pain. One of these aspects is the sensory discrimination of pain, in terms of location, intensity, duration, and quality. The main responsible of this discrimination is the thalamus, in particular the ventral posterior lateral (VPL) nucleus, and its projections to the primary somatosensory cortex (SI). Another important aspect is the affective or motivational, more related with the emotion that pain provokes in the individual who is suffering it (unpleasant feeling, fear, anxiety, and secondary autonomic reactions). In this case, the information travels by the spinoreticular and spinomesencephalic tracts, reaching several structures, such as the reticular formation, the superior colliculus, PAG, hypothalamus, and amygdala. In addition, another group of neurons constituting the anterior spinothalamic tract, reach the midline thalamic nuclei, that will connect with the anterior cingulate cortex and the insular cortex.

Apart from the anterolateral system, another fraction of information entering from the periphery travels along the dorsal columns, which will provide a more qualitative information of the stimulus that is reaching the nociceptive signaling system. This route is used essentially by mechanoreceptive afferents, which provide information about touch, vibration, and proprioception, and is known as the dorsal column-medial lemniscus system. This information travels directly by the central axons of primary sensory neurons in the dorsal columns ipsilateral to the site of entrance until the dorsal column nuclei in the medulla, where it decussates to reach the thalamus in the contralateral side and later on the cortex. This system also participates in the discriminative aspects of pain.

2.1.4 Descending Control of Pain

Once the nociceptive information arrives to the higher level centers, it is integrated in order to elicit a complex physiological response in front of the noxious stimuli, and also modulated in order to reduce the intensity of the painful sensation. The main mechanisms for pain modulation conform the descending pathway (Figure 2.3). One of the most important regions is the PAG in the midbrain, but there are other regions in the brainstem also involved in this process: parabrachial nucleus, medullary reticular formation, locus coeruleus, and raphe nuclei. These centers use noradrenaline, serotonin, dopamine, histamine, and acetylcholine to exert both excitatory and inhibitory effects on different sets of neurons in the dorsal horn. Then, they can act on synaptic terminals of nociceptive afferents, interneurons (excitatory and inhibitory), synaptic terminals of other descending pathways, and projection neurons. These contacts do not only act inhibiting the transmission of nociceptive information but also modulating it, as well as controlling the balance between excitation and inhibition in the spinal cord.

The main action of the PAG is to modulate nociceptive signaling in the dorsal horn by releasing endogenous opioids (encephalin, endorphins, and dinorphins) on the dendrites of nociceptive neurons and WDR neurons, causing hyperpolarization, thus inhibiting, of the second-order neurons. They

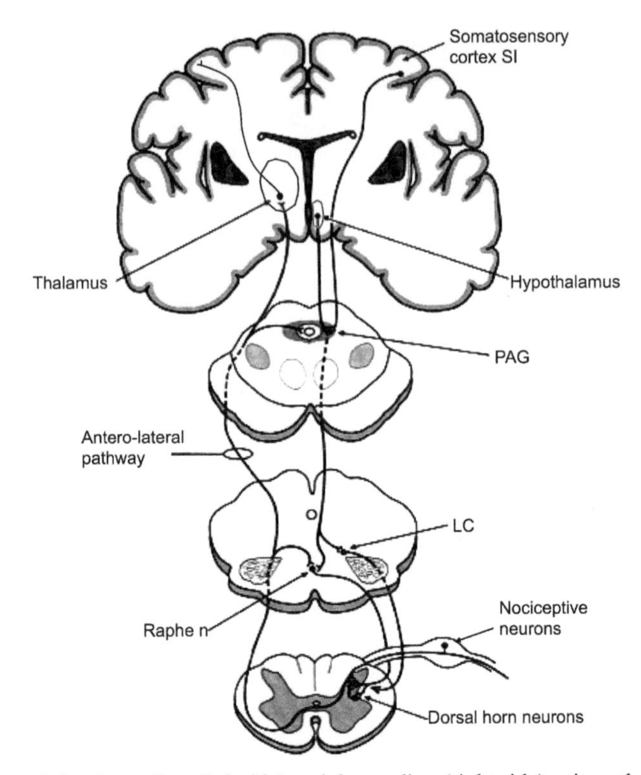

Figure 2.3 Ascending (left side) and descending (right side) pain pathways.

also release glycine on the terminals of primary afferents (A and C fibers), inducing presynaptic inhibition that reduces the release of neurotransmitters on the second-order neurons. Finally, the secretion of glutamate from the PAG excites the GABAergic interneurons in lamina II of the dorsal horn. This promotes the release of GABA on the second-order neurons, hyperpolarizing and therefore inhibiting them.

The PAG also causes depolarization of serotoninergic neurons in the raphe magnocellular nucleus (RMN), which project to second-order neurons in the dorsal horn via the raphespinal tract. The binding of serotonin to receptors 5-HT_1 and 5-HT_2 induces an increase in the conductance of potassium and therefore the hyperpolarization of the second-order nociceptive neurons. It also interacts with 5-HT_3 receptors in the dendrites of GABAergic interneurons in lamina II, inducing the release of GABA and the inhibition of second-order neurons. Something similar happens when PAG neurons secrete glutamate on the locus coeruleus neurons. Once depolarized, these neurons release noradrenaline, which causes hyperpolarization of the nociceptive second-order neurons by binding to α-adrenergic receptors.

Local circuits within the dorsal horn also play a role in modulating the nociceptive system. One of these systems was proposed by Melzack and Wall (1965), and called the "gate theory of pain," which actually is included under the afferent regulatory system of pain. This theory says that the activation of mechanoreceptors (large A fibers) can act on local interneurons to inhibit the transmission of information from C fibers to the dorsal horn projection neurons. This would explain how a mechanical stimulus such as scratching can temporarily give relief from pain in the same area.

Similarly, it has also been described a mechanism by which pain can inhibit pain. This phenomenon is called "diffuse noxious inhibitory control" (DNIC), or heterotopic noxious conditioning stimulation (HCNS) (Sprenger et al., 2011), and implies a spinal-medullary-spinal pathway. DNIC systems permit that a spinal neuron can be inhibited by a nociceptive stimulation applied in another part of the body (outside its receptive field), thus inhibiting the pain sensation after the application of a remote pain stimulation. WDR neurons and trigeminal nociceptive neurons play a key role in this phenomenon, which is regulated by serotoninergic pathways, and probably by opioids. DNIC effects are usually contralateral and extrasegmental, and highly dependent on the intensity of the stimulus. DNIC mechanisms may reflect alterations in the function of central descending inhibitory systems that could be potentially involved in chronic pain. In fact, research based on the use of DNIC has shown dysfunctions in this system in chronic pain conditions, such as fibromyalgia or irritable bowel syndrome (van Wijk and Veldhuijzen, 2010).

Other elements are also involved in pain regulation, such as the endogenous opioids. Several brainstem regions, most of them conforming the descending system of pain control, are susceptible to the action of these molecules, provoking an important analgesic effect. Endogenous opioids are classified in three groups, called enkephalins, endorphins, and dynorphins, which present different distribution along the nociceptive system. Enkephalins, for example, can be released by local neurons on the dorsal horn, then impeding the release of neurotransmitters from the terminals onto the projection neurons, and therefore diminishing their level of activity. This local circuit can also be the target of other descending inhibitory projections, therefore constituting a powerful control mechanism of the amount of nociceptive information able to reach superior centers. Endorphins are released in pain states within some brain regions, but they can also provide tonic analgesic effect in the dorsal horn. Dynorphins have been described to increase after neural injuries, and are related to the development of thermal hyperalgesia by acting on the NMDA receptors and driving to spinal sensitization (Ossipov et al., 2000).

2.2 Neurobiology of Neuropathic Pain

Neuropathic pain is defined as that pain provoked by a lesion or a dysfunction in the nervous system. Although sharing features with other kind of pain (inflammatory or cancer pain), neuropathic pain presents some particular characteristics. Nociceptive and inflammatory pain can be both symptoms of peripheral tissue injury, and present a clear defensive, beneficial component, whereas neuropathic pain is a symptom of neurological disease or injury, either affecting the peripheral or the central nervous system (Cervero, 2009), and instead of a defensive component it is considered as a maladaptive response.

Neuropathic pain states are characterized by an almost complete lack of correlation between the intensity of peripheral noxious stimuli and of pain sensation, and are produced by neurological lesions that cause abnormal impulse activity generated in nerve sprouts, neuromas, or dorsal root ganglion cells, ephaptic coupling between adjacent nerve fibers and abnormal responses of peripheral nociceptors and CNS neurons. Neuropathic pain syndromes produce pain sensations well outside the range of the sensations produced by the normal nociceptive system, even after serious peripheral injury or inflammation (Cervero, 2009). Neuropathic pain may include spontaneous, or stimulus-independent pain that has been described as shooting, burning, lancinating, prickling, and electrical. Evoked, or stimulus-dependent, neuropathic pain includes allodynia, defined by the IASP as "pain due to a stimulus which does not normally provoke pain." These stimuli may be nonnoxious heat, light touch, or even a puff of cool air. Moreover, mechanical allodynia may be static, as evoked by light touch, or dynamic, as evoked by a light brush of the skin. Hyperalgesia is identified when a stimulus that normally produces nociceptive responses produces exaggerated responses.

Epidemiological studies on the prevalence of neuropathic pain indicate a high incidence of about 5% of the general population (Bouhassira et al., 2008; Dieleman et al., 2008). Associated risk factors include gender, age, and

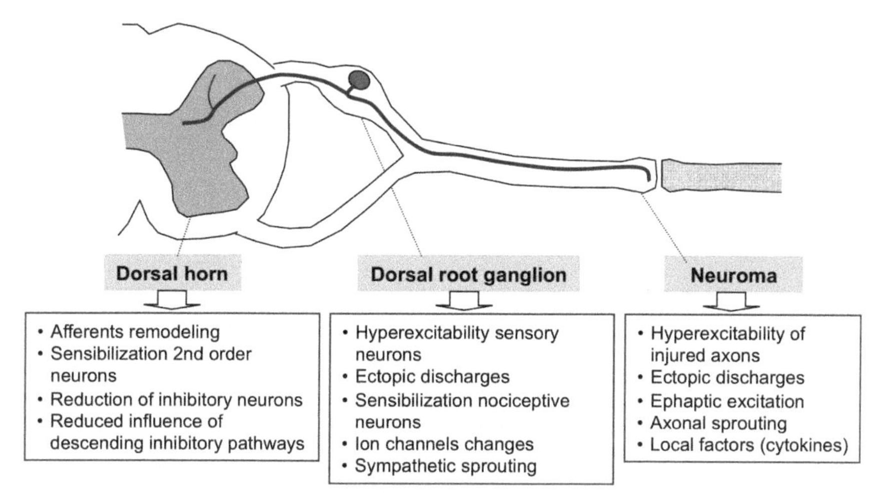

Figure 2.4 Schematic representation of the main peripheral mechanisms involved in the generation of neuropathic pain after a peripheral nerve injury.

anatomical site of the injury. Emotional and cognitive factors influence how patients react to chronic pain (Haythornthwaite et al., 2003), but it is much less certain if these factors contribute to the risk of developing pain.

The different forms through which neuropathic pain manifests suggest that different mechanisms are likely to mediate the different features of the condition. Features that appear most often related to neuropathic pain include tingling or numbness, pain evoked by heat or cold, and sensation of heat or a burning-like quality.

2.2.1 Mechanisms of Neuropathic Pain

Five key processes are considered in the neurobiological approach to the mechanisms of pain and hyperalgesia following a peripheral nerve injury (Figure 2.4): (1) the process of nociceptor activation and sensitization, responsible for the initial signaling of injury and the peripheral changes in the nociceptive system induced by a noxious stimulus; (2) the process of central amplification of nociceptive signals, known as central sensitization, generated by synaptic strengthening of connections between CNS neurons and responsible for the enhanced excitability that accompanies persistent pain states; (3) the process whereby activity in low-threshold sensory receptors from undamaged peripheral areas can access the nociceptive system and evoke pain sensations and hyperalgesic states (e.g., touch-evoked pain, tactile allodynia); (4) the loss of endogenous inhibition produced at spinal level paired with changes in descending inhibitory pathways; and (5) neuroplastic changes at subcortical and cortical levels (Navarro et al., 2007).

2.2.2 Nerve Injury-induced Changes in Transduction

Peripheral nerve injury leads to a redistribution of transducers in the wrong place within a neuron, and this event carries two consequences: the emergence of mechanical sensitivity at sites that are normally mechanically insensitive and mechanical allodynia. Even mechanical stimuli associated with physiological functions, such as movement of tissue associated with blood flow, may also stimulate these transducers.

Normal pain sensations are normally elicited by activity in unmyelinated (C) and thinly myelinated (Aδ) primary afferent neurons. These nociceptors are usually silent in the absence of stimulation, and respond best to stimuli that are potentially noxious. In neuropathic pain disease, after a peripheral nerve lesion, axotomized neurons become abnormally sensitive and develop pathological spontaneous activity generated at any anatomical level proximal to the lesion. These pathological changes are underpinned by dramatic molecular and cellular changes at the primary afferent nociceptor that are triggered by the nerve lesion.

A long time ago, spontaneous ectopic activity was demonstrated in awakened human amputees with phantom limb pain, by microneurographic single-fiber recordings from afferent fibers projecting into the neuroma, as well as barrages of action potential firing. After a nerve injury there is an increase in membrane excitability. Spontaneous discharges in DRG neurons have been recorded from cells of both intact and injured nerves. There are phenotypic changes in injured neurons but also in uninjured ones, driven by cytokines and growth factors released by denervated Schwann cells.

An embryonic channel, Nav1.3, is upregulated in damaged peripheral nerves, and this is thought to promote ectopic spontaneous activity in primary afferent neurons. Also, genes for the voltage-gated sodium channels Nav1.8 and Nav1.9 are expressed selectively in nociceptive primary afferent neurons. These fibers acquire a unique sodium-channel expression profile after nerve injury, with upregulation of Nav1.3 and downregulation of Nav1.7, Nav1.8, and Nav1.9. Changes of these channels are responsible for the lowering of the action-potential threshold and consequent hyperexcitability, playing an important role in the genesis of neuropathic pain (Omana-Zapata et al., 1997).

Entrance of calcium ions into the nerve endings regulates growth-related proteins and can participate also in the increased excitability of the injured neurons. In vitro, N and L-type calcium channels have been found to contribute to CGRP release from injured nerves. Blockade of N-, T-, and P-type calcium channels reduces neuropathic pain.

Related to thermal stimuli, normal body temperature can elicit spontaneous activity of primary afferents as a result of a change in the activation threshold of the noxious heat-sensitive TRPV1 channel. Damage to peripheral nerves provokes upregulation of TRPV1 that are located predominantly on uninjured C fibers and A fibers. These changes might contribute to the development of C-nociceptor sensitization and the associated symptom of heat hyperalgesia.

In transduction for mechanical stimuli, ASICs seem to be involved in static mechanical hyperalgesia. Nerve lesion fibers developing adrenergic sensitivity. It is known that in amputees, the perineuronal administration of norepinephrine induced intense pain. Neuroma after injury has both afferent C fibers and efferent post-ganglionic sympathetic C fibers which release noradrenaline and adrenaline. With high sympathetic activity there is a raised sensitivity of the regenerating sprouts toward the detection of nociceptive substances, such as bradykinin, serotonin, histamine, and capsaicin evoking depolarization and ectopic firing. Afferent excitability can also be increased by the combination of a downregulation of inhibitor transducers, such as opioid receptors, and the upregulation of excitatory transducers, such as the ATP receptor P2X3.

After Wallerian degeneration, products such as nerve growth factor (NGF) are released in the vicinity of spared fibers triggering the release of tumor-necrosis factor-α (TNF-α), channels and receptors expression (sodium channels, TRPV1 receptors, adrenoceptors...) thereby altering also the properties of uninjured afferents (Wu et al., 2002). Research is more focused now on the variety of changes that might occur in uninjured axons, as these neurons are still connected with their peripheral organs and could have a pivotal role in the generation of neuropathic pain.

2.2.3 Central Sensitization

Neuropathic pain may arise either as a result of peripheral sensitization of intact afferents or due to central sensitization. Central sensitization has been defined as "a prolonged but reversible increase in the excitability and synaptic efficacy of neurons in central nociceptive pathways." The enhanced synaptic transmission is manifested by long-term potentiation (LTP) following a short train of stimulation of C fibers. The transition of LTP between spinal interneurons involves glutamate and neurokinin 1 receptors.

The function of nociceptive pathways is increased due to high membrane excitability, synaptic efficacy, and reduced central inhibition. The net effect of central sensitization is to recruit previously subthreshold synaptic inputs to nociceptive neurons, generating an increased or augmented action potential output: a state of facilitation, potentiation, augmentation, or amplification (Latremoliere and Woolf, 2009). It is manifested as allodynia (touch-evoked pain), enhanced temporal summation (escalating pain in response to a constant stimulus), hyperalgesia (exaggerated pain experience to a standardized noxious stimulus), and secondary hyperalgesia (pain and hypersensitivity beyond the dermatome of the nerve injury).

Primary nociceptive neurons release glutamate, SP, CGRP, and ATP as neurotransmitters and neuromodulators. After injury, hyperexcitability is established with greater release of neurotransmitters in the spinal cord. They interact with NMDA, AMPA, mGluR, NK1R, and P2X receptors causing depolarization of nociceptive second-order neurons and scattering throughout the spinothalamic pathway.

Peripheral nerve injury leads to an increase in the general excitability of wide dynamic range neurons in the dorsal horn with multiple synaptic inputs from nociceptive and non-nociceptive system. This central sensitization is initiated and maintained by activity in pathologically sensitized C fibers. After peripheral nerve injury there are presynaptic functional changes that increase synaptic strength: the synthesis of transmitters and neuromodulators and more calcium. Aβ touch fibers express increased levels of neuropeptides, such as CGRP and SP, and enhance activity of excitatory amino acid transmission via NMDA receptors.

Healthy nerve terminals uptake NGF and other growth factors from their target cells and transmit them by axonal transport to the DRG neurons. After nerve transection, this growth factors supply is interrupted, so that gene transcription and protein synthesis are altered. At the level of transcription control in the DRG neurons, the c-jun gene can be inducted 1 day after axotomy. It is well known that c-Jun expression in the DRG neurons after nerve transection is associated with changes in neuropeptide levels: SP and CGRP decrease; galanin and NO synthase (NOS) increase dramatically during weeks following axotomy. The induction of c-jun of axotomized neurons has been closely related with inhibitory transynaptic neuron death or apoptosis by NGF starvation.

The increased release and production of NOS at the intraspinal presynaptic terminal may facilitate afferent synaptic transmission to the dorsal horn neurons contributing to spinal neuronal sensitization and hyperalgesia. Repetitive noxious stimulation leads to the increased activity of NMDA and AMPA receptors, which produce an influx of extracellular Ca^{2+} and activation of PKC in dorsal horn neurons. The increased intracellular Ca^{2+} induces the expression of c-fos. Fos protein is involved in the transcriptional control of genes that encode a variety of neuropeptides, including enkephalin and dynorphin. Enkephalin typically produces antinociceptive effects. Dynorphin has direct excitatory effects on spinal projection neurons and may also produce inhibition via a negative feedback mechanism on dynorphin-containing neurons. The net effect of these changes may have complex modulations in the development of central plasticity.

Postsynaptically, second-order dorsal horn neurons also abnormally express Nav1.3 after peripheral nerve injury. Physiologically, dorsal horn neurons receive a strong inhibitory control by GABA releasing interneurons that are lost by apoptosis after partial peripheral nerve injury, thus favoring central sensitization. Other postsynaptic changes involve phosphorylation of NMDA subunits and increased receptor density due to trafficking and enhanced synthesis of ion channels and scaffold proteins. These changes underlying central sensitization occur in the dorsal horn, amygdala, anterior cingulate gyrus, and prefrontal cortex.

Continuous and sustained afferent inputs, due to hyperactivity of the damaged nociceptive fibers, into the spinal cord cause a state of spinal sensitization. This is related to the demonstrated phenomenon of wind-up, in which noxious stimuli applied to the skin also enhance the excitability of dorsal horn units, producing exaggerated responses to subsequent stimuli (Mendell and Walsh, 1965). Another mechanism of intraspinal disinhibition following peripheral nerve injury involves a trans-synaptic reduction in the expression of the potassium-chloride cotransporter KCC2 in lamina I neurons, which disrupts anion homeostasis in these neurons (Mòdol et al., 2014). The effect is that GABA release from normally inhibitory interneurons and now exerts an excitatory action increasing central sensitization. Dorsal horn neurons receive a powerful facilitatory but mostly inhibitory descending modulating control from supraspinal brainstem centers. A loss of function in descending inhibitory serotonergic and noradrenergic pathways contributes to central sensitization and pain chronification. Peripheral nerve injury activates spinal cord glia and causes these cells to enhance pain by releasing proinflammatory cytokines and glutamate producing also central sensitization.

2.2.4 Low-threshold Aβ Fiber-mediated Pain

Neuropathic pain involves a profound switch in sensitivity such that low-threshold Aβ fibers, which normally signal innocuous sensations, may begin after neural lesions to produce pain sensation (Witting et al., 2006). A number of changes either independently or combined can promote Aβ fiber-mediated pain: central sensitization, disinhibition, and central afferent terminal sprouting.

Differential block induced by compression of the radial nerve in patients with nerve injury or exposed to experimental pain clearly demonstrated that pain induced by light brushing was mediated through Aβ fibers, whereas thermal pain was mediated through unmyelinated C fibers. Intrathecal morphine reversed nerve injury-induced thermal, but not tactile, hypersensitivity.

After peripheral nerve injury the most spontaneously active fibers are Aβ and Aδ fibers (>80%), whereas C fibers represent a small (0–30%) portion of the active population. This hypersensitivity includes areas outside the injured nerve territories and occurs largely in the absence of peripheral sensitization. It is typically associated with a loss of C fiber peripheral terminals and disappears when conduction in large myelinated fibers is blocked. After nerve injury polysynaptic and monosynaptic Aβ fiber inputs in the most superficial laminae I and II of the dorsal horn are increased. These laminae normally receive only input from Aδ and C fibers. After an injury, this clear organization can be lost, as some Aβ fibers arriving to lamina III–IV can produce aberrant sprouting and reach outer laminae. This may imply that some innocuous, tactile information will be processed abnormally in a nociceptive territory, constituting a potential mechanism for allodynia (Costigan and Woolf, 2000).

2.2.5 Changes in Endogenous Inhibitory Pathways, Disinhibition, and Plasticity

After nerve injury there is loss of pre- and postsynaptic inhibition in the spinal cord. This can occur because of death of inhibitory interneurons caused by the excitotoxicity of the lesion, reduction in the release of inhibitory neurotransmitters from surviving interneurons, and reduction in the expression of inhibitory transmitter receptors. In nociceptive lamina I neurons, the transmembrane gradient for chloride ions changes. GABA receptors no longer lead to hyperpolarization, instead depolarization is induced, provoking excitation and spontaneous activity (Keller et al., 2007). Independently, there is loss of GABAergic interneurons by apoptosis compromising the dorsal inhibition.

After GABAergic or glycinergic blockade or removal, tactile allodynia is induced and synaptic currents from Aβ fibers to nociceptive lamina I neurons increase.

The tonic noradrenergic inhibition that acts on α2-adrenoceptors appears to be suspended after lesion thus, the net effect of descending adrenergic input goes paradoxically, from inhibition to facilitation. Other descending pathway in the modulation of the nociceptive input are the μ opioid receptors, which reduce their expression together with less sensitivity of dorsal horn neurons to its agonists after nerve injury.

Activated microglia synthesize and release prostaglandins, chemokines, NOS, and cytokines (TNFα, IL1, IL6...) acting as chemical mediators to amplify the microglial reactivity favoring the elevation of these molecules in the dorsal horn. Reactive microglial cells are also responsible for the release of cathepsine-S protease that causes the proteolysis of a transmembrane glycoprotein called fractalkine (CX3CL1) which interacts with its receptor (CX3XR1) located in the reactive microglial cell membrane, maintaining its reactivity and therefore contributing to the preservation of neuropathic pain. Microglia can also provoke neuronal death by producing ROS, proapoptotic cytokines and by a diminished glutamate uptake. Astrocytes also become activated after peripheral nerve injury with a slower onset and more prolonged time course than microglia, but also playing a role in the maintenance of neuropathic pain hypersensitivity.

The energy depletion in the injured neurons decreases the intracellular ATP concentration and consequent K_{ATP} channels activation. The activation causes potassium ions outflow leading to hyperpolarization, less excitability, and reduction in the release of neurotransmitters. However, K_{ATP} expression is decreased after peripheral nerve lesion of myelinated Aδ nociceptive fibers enhancing hyperexcitability.

In the thalamus, there is an upregulation of nicotinic and cannabinoid receptors after peripheral nerve injury, suggesting that supraspinal receptors in the thalamus may contribute to the modulation of neuropathic pain responses. On the other hand, μ-opioid receptor mediated G-protein activity is reduced producing desensitization in receptors from this region. Also, NKCC1 and KCC2 dysregulation in the spinothalamic pathway is produced after sciatic nerve section, suggesting that neuropathic pain is maintained by reducing inhibitory inputs not only in the spinal cord but also at thalamus and cortex (Mòdol et al., 2014).

In summary, multiple and different mechanisms operate after nerve injury to increase excitability and reduce inhibition in the central pain pathway.

Apart from the loss of descending inhibition, another feature contributing to the hyperexcitablity in the pain pathway after injury is disinhibition. Silent circuits and synapses in normal conditions can become activated after peripheral nerve injuries. The disinhibition can be produced by a shift in the properties of inhibitory receptors, for example, the loss of inhibitory function in GABA receptors, in relation with a shift in the function of chloride transporters NKCC1 and KCC2 (Mòdol et al., 2014).

Neural plasticity occurring after neuronal lesions is also detectable in reflex circuits, which are usually used as an indirect measure of central hyperexcitability. Electrophysiological changes caused by the lesion and following plastic reorganization can produce the appearance of hyperreflexia and an increase in wind-up responses that can eventually lead to spasticity and neuropathic pain (Valero-Cabré and Navarro, 2010; Redondo-Castro et al., 2010).

2.2.6 Changes in Subcortical and Cortical Regions

Some groups of brainstem neurons are related with nociceptive modulation, forming the called "brainstem pain modulating system." It includes the midline PAG-RVM system, the more lateral and caudal dorsal reticular nucleus (DRt) and caudal ventrolateral medulla (cVLM). The descending projection of RVM includes the nucleus raphe magnus and the adjacent reticular formation. Serotonin and noradrenaline are the main neurotransmitters in these descending pathways. The predominant source of serotonergic input to the spinal cord arises from the nucleus raphe magnus. Serotonin causes hyperpolarization of afferent nociceptive fiber terminals and dorsal horn projection and it produces excitation in spinal GABAergic interneurons. Noradrenaline causes hyperpolarization of projection neurons and over terminals of primary afferent fibers inducing excitation of dorsal horn inhibitory interneurons. There are two types of neurons in the PAG-RVM system: "ON" cells facilitate and "OFF" cells inhibit pain transmission (Fields et al., 2000).

Neuropathic pain induces hyperexcitation of specific nociceptive and WDR neurons in the spinal cord causing a potentiation of the "ON" neurons response and a decrease of the "OFF" neurons. The preferential activation of "ON" cells located in RVM causes hyperalgesia, whereas hypoalgesia is achieved by the activation of RVM "OFF" cells. A sensitization of "ON" RVM neurons is also induced by overexpression of NMDA/AMPA, Trk-B, and NK1 receptors, whereas μ opioid receptor expression decreases. Under these circumstances, "ON" RVM neurons do not respond to inhibitory signals from PAG, whereas they are highly stimulated by ascending inputs that release glutamate, SP, and dynorphin over thalamic and brainstem neurons. PAG neurons are also sensitized due to overexpression of several receptors including NMDA/AMPA and SP/NK1 but also overexpression of glutamate and BDNF. These PAG-BDNF-positive neurons project their axons over the "ON" RVM neurons, enhancing their depolarization via Trk-B and NMDA/AMPA receptors.

Furthermore, "OFF" RVM neurons mainly express NMDA/AMPA and TRPV1 receptors. After neuropathic and/or inflammatory painful insults, there are molecular changes in "OFF" RVM neurons such as an overexpression of GABA-A and kappa opioid receptors provoking hyperpolarization and reduction of their antinociceptive effect on spinal cord dorsal horn neurons. The increase of microglial and astroglial reactivity in RVM also contributes to the misbalance of the normally inhibitory descending pathways, by releasing several mediators that facilitate the excitation of "ON" RVM neurons and their excitatory effect on dorsal horn neurons (Wei et al., 2008).

A simple cutaneous nerve anesthetic blockade causes fast and reversible reorganization of dorsal column nuclei and thalamus. In primates, immediately after section of ulnar and median nerves there is a change in hand representation mapping in their brainstem cuneate nucleus, maintained over several days. Reorganization in cortex and thalamus territories is delayed a few weeks to months, depending on the growth of new connections and the expansion of thalamic receptive fields that tend to be larger than in somatosensory cortex. Cortical and subcortical reorganization and plasticity depend on the abnormal projections from peripheral and spinal levels.

Peripheral nerve injuries result in loss of evoked activity by deafferentation in the corresponding cortical map. Then, plastic changes occur resulting in reduction of the cortical area representing the denervated part of the body, in favor of expansion of adjacent representations from intact sources with hardly discernible somatotopy. The brain plasticity is not very accurate per se, as it was demonstrated in several cases of nerve section and repair in humans where sensory mislocalizations persist for many years due to misdirected reinnervation. The reorganization has different time courses depending on the severity of the lesion. After experimental sensory deafferentation of one finger in a human patient, the expansion of cortical representations of intact fingers is very fast and is recovered in minutes when the sensory blockade is stopped. On the other hand, amputees suffer a slower brain plasticity process for a few weeks to become permanent with reduction of affected area and expansion of adjacent cortical regions. The basis of phantom sensations secondary to limb amputation seems to include reorganization phenomena at the cortical and subcortical levels. Furthermore, a sensory map corresponding to the different portions of the missing limb can be traced in the stump or in the face of some amputee subjects (Ramachandran and Hirstein, 1998).

The first changes in the brain produced by a peripheral nerve injury involve changes in synaptic efficacy, including increased excitatory neurotransmitter release, increased density of postsynaptic receptors, changes in membrane conductance, or removal of inhibitory projection resulting in unmasking of inactive synaptic connections at cortical and subcortical levels. Loss of GABAergic inhibition is the main cause of the short-term plastic changes in the cortex. In the same way that occurs in the spinal cord, there is an upregulation of Nav1.3 in the third-order nociceptive neurons in the thalamus resulting in hyperexcitability and expanded receptive fields. In situations of chronic deafferentation, structural mechanisms like LTP or LTD phenomena, sprouting for the formation of *de novo* connections and synaptogenesis can be reduced by NMDA receptor antagonist administration at the beginning of the plastic process.

At subcortical levels, transneuronal atrophy associated with retraction of axons and compensatory axonal sprouting seems to play a significant influence on the reorganization of somatotopic maps in the brain cortex. Plastic reorganization after nerve injuries has been related with structural changes in dendritic arborization within the cortex. Brain reorganization processes in adult subjects seem to occur primarily through changes in the strength and efficacy of existing synapses, rather than implicate active remodeling of connections.

In summary, neuropathic pain triggers plastic changes in the descending pain modulatory pathway: "ON" cell activation and "OFF" cell inactivation from the PAG-RVM system, to increase pain facilitation in the spinal cord. In addition, reorganization of ascending projections takes place sequentially from the dorsal horn to the brainstem, the thalamus and finally the somatosensory cortex.

References

Bouhassira, D., Lanteri-Minet, M., Attal, N., Laurent, B. and Touboul, C. (2008). Prevalence of chronic pain with neuropathic characteristics in the general population. Pain 136:380–7.

Cervero, F. (2009). Spinal cord hyperexcitability and its role in pain and hyperalgesia. Experimental Brain Research 196:129–37.

Costigan, M. and Woolf, C. J. (2000). Pain: Molecular mechanisms. Pain 1:35–44.

Dieleman, J. P., Kerklaan, J., Huygen, F. J., Bouma, P. A. and Sturkenboom, M. C. (2008). Incidence rates and treatment of neuropathic pain conditions in the general population. Pain 137:681–8.

Fields, H. L. (2000). Pain modulation: Expectation, opioid analgesia and virtual pain. Prog Brain Res 122:245–253.

Haythornthwaite, J. A., Clark, M. R., Pappagallo, M. and Raja, S. N. (2003). Pain coping strategies play a role in the persistence of pain in post-herpetic neuralgia. Pain 106:453–60.

Keller, A. F., Beggs, S., Salter, M. W. and De Koninck, Y. (2007). Transformation of the output of spinal lamina I neurons after nerve injury and microglia stimulation underlying neuropathic pain. Mol Pain 3:27.

Latremoliere, A. and Woolf, C. J. (2009). Central sensitization: A generator of pain hypersensitivity by central neural plasticity. J Pain 10:895–926.

Melzack, R. and Wall, P. D. (1965). Pain mechanisms: A new theory. Science 150:971–9.

Mendell, L. M. and Walsh, J. H. (1965). Responses of dorsal cord cells to peripheral cutaneous unmyelinated fibres. Nature 206:97–9.

Mòdol, L., Cobianchi, S. and Navarro, X. (2014). Prevention of NKCC1 phosphorylation avoids down-regulation of KCC2 in central sensory pathways and reduces neuropathic pain after peripheral nerve injury. Pain 155:1577–90.

Navarro, X., Vivó, M. and Valero-Cabré, A. (2007). Neural plasticity after peripheral nerve injury and regeneration. Progr Neurobiol 82:163–201.

Omana-Zapata, I., Khabbaz, M. A., Hunter, J. C., Clarke, D. E. and Bley, K. R. (1997). Tetrodotoxin inhibits neuropathic ectopic activity in neuromas, dorsal root ganglia and dorsal horn neurons. Pain 72:41–9.

Ossipov, M. H., Lai, J., Malan, T. P. and Porreca, F. (2000). Spinal and supraspinal mechanisms of neuropathic pain. Ann N Y Acad Sci 909:12–24.

Ramachandran, V. S. and Hirstein, W. (1998). The perception of phantom limbs. The D. O. Hebb lecture. Brain 121:1603–30.

Redondo-Castro, E., Udina, E., Verdú, E. and Navarro, X. (2011). Longitudinal study of wind-up responses after graded spinal cord injuries in the adult rat. Restor Neurol Neurosci 29:115–26.

Sprenger, C., Bingel, U. and Büchel, C. (2011). Treating pain with pain: Supraspinal mechanisms of endogenous analgesia elicited by heterotopic noxious conditioning stimulation. Pain 152:428–39.

Valero-Cabré A. and Navarro, X. (2002). Changes in crossed spinal reflexes after peripheral nerve injury and repair. J Neurophysiol 87:1763–71.

van Wijk, G. and Veldhuijzen, D. S. (2010). Perspective on diffuse noxious inhibitory controls as a model of endogenous pain modulation in clinical pain syndromes. J Pain 11:408–19.

Wei, F., Guo, W., Zou, S., Ren, K. and Dubner, R. (2008). Supraspinal glial-neuronal interactions contribute to descending pain facilitation. J Neurosci 28:10482–95.

Witting, N., Kupers, R. C., Svensson, P. and Jensen, T. S. (2006). A PET activation study of brush-evoked allodynia in patients with nerve injury pain. Pain 120:145–54.

Woolf, C. J. and Ma, Q. (2007). Nociceptors – Noxious stimulus detectors. Neuron 55:353–64.

Wu, G., Ringkamp, M., Murinson, B. B., Pogatzki, E. M., Hartke, T. V., Weerahandi, H. M., Campbell, J. N., Griffin, J. W. and Meyer, R. A. (2002). Degeneration of myelinated efferent fibers induces spontaneous activity in uninjured C-fiber afferents. J Neurosci 22:7746–53.

3

The TIME Implantable Nerve Electrode

Tim Boretius[1,2] and Thomas Stieglitz[2,3,4,*]

[1]Neuroloop GmbH, Freiburg, Germany
[2]Laboratory for Biomedical Microsystems, Department of Microsystems Engineering-IMTEK, Albert Ludwig University of Freiburg, Freiburg, Germany
[3]BrainLinks-BrainTools, Albert Ludwig University of Freiburg, Freiburg, Germany
[4]Bernstein Center Freiburg, Albert Ludwig University of Freiburg, Freiburg, Germany
E-mail: stieglitz@imtek.uni-freiburg.de
*Corresponding Author

Transversal intrafascicular multichannel electrodes (TIMEs) have been developed using state-of-the-art micromachining technologies. This chapter presents the different designs of the implantable thin film electrode array, cable and connector development, and the assembling of components into implantable devices. As a last step towards translational research and human clinical trials, implantable devices have been fabricated under a quality management system to meet the essential requirements of active implantable medical device regulations.

3.1 Introduction

Research on interfaces to the nervous system is driven by application needs and knowledge about the performance, effects, and side effects of existing devices. Knowledge from material sciences and "novel" manufacturing technologies that go beyond the integration level and degree of miniaturization of precision mechanics propose regularly new approaches to obtain better spatial resolution with less foreign body reaction and more and more recording or stimulation channels. Promises have been made in many cases but only few devices have been transferred into chronical applications and even fewer in humans in clinical practice.

Based on our expertise in microsystems engineering and promising results published about the performance of longitudinal interfascicular electrodes made with means of precision mechanics as well as micromachined ones, we started the endeavor to develop a new type of intrafascicular electrode that should deliver good levels of special selectivity on a fascicular or even subfascicular level. Investigations should not only include the technical development but also target a first-in-man clinical trial. Therefore, designs had to be developed for small and large animal models to prove that selective stimulation of peripheral nerves with little crosstalk at low stimulation thresholds is feasible. From the engineering side, system integration with cables and connectors to stimulators and percutaneous wires for human clinical trials had to be developed. Devices had to be implanted and prove biocompatibility and functionality for up to 30 days as legally defined period for a first-in-man study. Nevertheless, questions on long-term functionality were investigated with the objective to come to much longer stability of implanted microsystems. In the course of the developments, several designs were made for small and large nerves to finally come to a version for our human subject. This last design of TIME probes had to be manufactured under quality management requirements and regulations to fulfill "essential requirements" according to medical device laws in Europe. This chapter is a journey through the different steps of TIME developments and will not only deliver the final results but show how iterative research work helps to improve device functionality and performance when only little pre-existent knowledge is available.

3.2 Design and Development of TIME Devices

Previous experience of other research groups in interfaces to the peripheral nervous system showed well that invasiveness and selectivity have to be well balanced with respect to the envisioned medical application. The goal of treating phantom limb pain after amputation trauma of a hand by implanting electrodes for electrical stimulation of sensory feedback required coverage of the whole arm nerve(s) with selectivity on the (sub-)fascicular level. While cuff electrodes deliver good selectivity to superficial fibers and longitudinal intrafascicular electrodes (LIFE) are superior with local selectivity but lack cross-sectional coverage, we propose the transversal intrafascicular multichannel electrode (TIME) to get both nerve cross-sectional coverage and local selectivity in one device (Figure 3.1) (Boretius 2009).

Good experience of implanting thin-polymer-based electrodes longitudinally in the nerve (tfLIFE) brought us the idea to try out a modified design to have similar foreign body reaction and selectivity of the adjacent fascicles and to cover the whole cross section of the nerve by transversal implantation.

This idea led to the first basic electrode design of a transversal intrafascicular multichannel electrode which is still similar to the tfLIFEs. An electrode arrangement comprising three active stimulation sites plus an indifferent and a ground electrode was designed on a substrate that subsequently becomes kinked within the center line and gets inserted into the peripheral nerve. This kinking allows to have electrode sites on both sides of the implant without sophisticated (non-standard) micromachining technologies that quite often leads to low yield and prevents medium-scale manufacturing of devices (in our case tens to hundreds) with a comparable quality at early stages when technology and manufacturing readiness level are still low. In addition to the transversal implantation procedure, it was hypothesized that a corrugated substrate allows better adaptation to the different fascicles and thus lower stimulation thresholds, less crosstalk, and higher selectivity. The engineering challenge was the development of a post-micromachining technology to achieve a corrugation of polymer substrates with integrated electrode sites (see below).

An estimation of the forces at the material–tissue interface due to lateral displacement of the implant has been calculated with the envisioned geometrical design parameters and the Young's modulus of the substrate material (polyimide, UBE U-Varnish-S) that had been determined as $\mathrm{E} = 7$ GPa in another study (Rubehn 2010). We assumed the interconnect-electrode structure as a rectangular beam with a one side restraint and a connector side and neglected the thin platinum metallization in the first approximation. The beam had a thickness of $\mathrm{h} = 11\ \mu\mathrm{m}$, a width of $\mathrm{b} = 580\ \mu\mathrm{m}$, and a length of $\mathrm{w} = 3400\ \mu\mathrm{m}$:

$$I_z = \frac{bh^3}{12} = 6.43 \cdot 10^{-20} m^4 \tag{3.1}$$

$$I_y = \frac{b^3h}{12} = 1.79 \cdot 10^{-16} m^4 \tag{3.2}$$

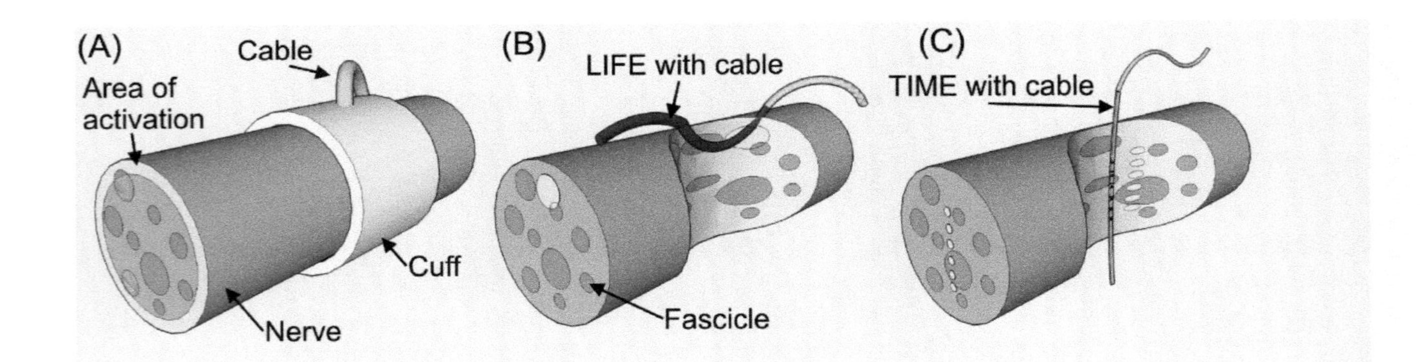

Figure 3.1 Selectivity of different electrode types (activated nerve region in white): Cuff electrodes activate the superficial perimeter of the nerve first (left); LIFEs (middle) have very local activation properties; TIME allows multiple small areas of activation over the nerve cross section.

Source: Reprinted from Boretius, T., Badia, J., Pascual-Font, A., Schuettler, M., Navarro, X., Yoshida, K. and Stieglitz, T. A transversal intrafascicular multichannel electrode (TIME) to interface with the peripheral nerve, *Biosens Bioelectron*, 26(1): 62–69. Copyright (2010), with permission from Elsevier.

The given geometry leads to moments of inertia in the direction of the structure I_z and the structure orthogonal to it I_y.

The application of a point load on one side leads to a displacement:

$$w(x) = \frac{F}{6EI} \cdot \left(x^3 - 3lx^2\right) \tag{3.3}$$

$$w(x = l) = \frac{l^3}{3}\frac{F}{EI} \tag{3.4}$$

A displacement of 10 μm inserted into the beam equation (Equation 3.3) at the end of the structure at x = 1 (Equation 3.4) leads to a force of F = 0.343 μN in the normal direction and a force of F = 956 μN in the orthogonal direction with a relationship of the moments of inertia of I_y: I_z = 2783.

Since the first proof-of-concept studies in vivo were scheduled with rat animal models, the design was adapted to the anatomy of the rat sciatic nerve which has a diameter of approximately 1.2 mm. The electrodes were designed with a final pitch after corrugation of 300 μm (400 μm before), thus ensuring a proper positioning and contacting of the three fascicles within the rat sciatic nerve. The electrode sites had an effective diameter of 60 μm (80 μm metallization) and the electrode tracks had a width of 50 μm (Figure 3.2).

The design of this first version of a TIME device had an overall width of 580 μm, a thickness of approximately 12 μm, and a length from center to interconnection pads of 3.4 cm. The fins in the device center are used for positioning purposes within the moldform and will be cut after corrugation (for corrugation see section below).

Metallization had been performed by sputter depositing of platinum and lift-off structuring technique, so far. Limited charge injection capacity and foreign body reactions around the implant limited the performance of the first thin-film LIFE (tfLIFE) in a human to 10 days. To enhance the performance of the device, that is, decreasing the impedance and cut-off frequency (as well as increasing the charge injection capacity), different coatings for the electrode sites were investigated. Details of process and coating technology are given in detail in the following sections.

3.2.1 Process Technology to Manufacture TIMEs

The electrode manufacturing was conducted in a class 1000 cleanroom using state-of-the-art micromachining processes (Figure 3.3).

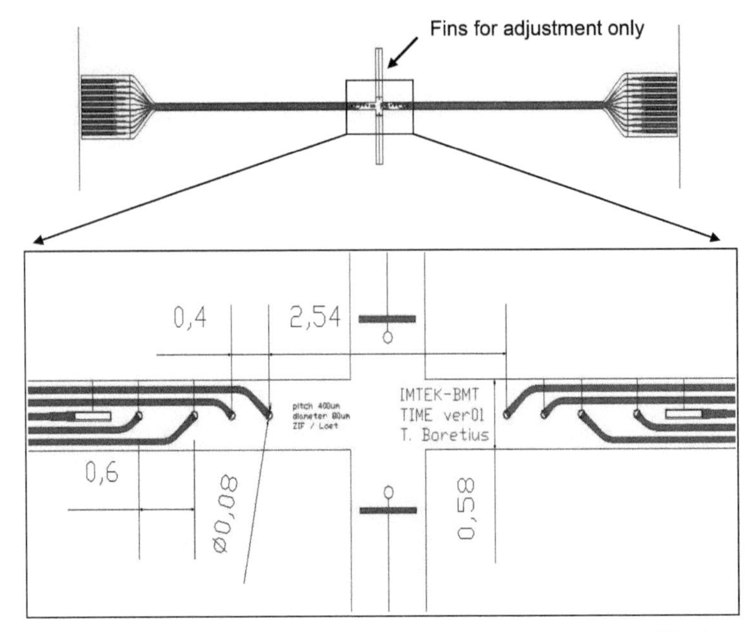

Figure 3.2 Design of the first version of a TIME. Red color indicates metallization and blue color indicates polyimide borders. All dimensions are presented in millimeters.

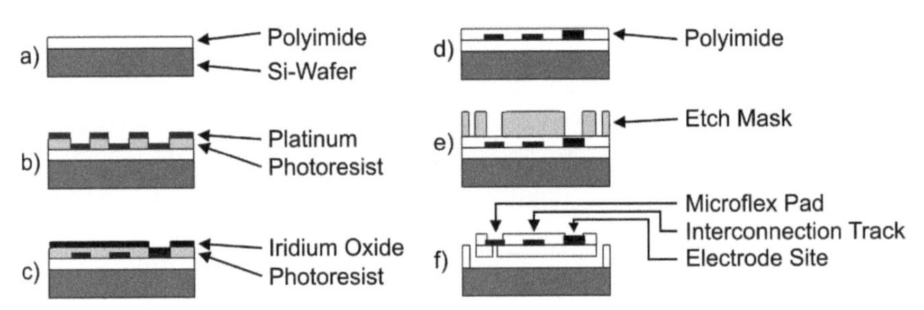

Figure 3.3 Schematic view of the micromachining process to manufacture TIME devices (see text for details).

First, a 5.5-μm-thick layer of polyimide (U-Varnish-S, UBE Industries, Tokyo, Japan) was spin coated onto a silicon wafer that was solely taken for handling purposes (Figure 3.3a). A layer of lift-off resist was deposited and patterned by photolithography using a mask aligner (MA/BA6, Carl Zeiss AG, Jena, Germany). Next to an oxygen flash of 30 seconds that enhances the adhesion of metal to polymer, a 300-nm-thick layer of platinum was deposited by sputtering technique (Leybold Univex 500, Leybold GmbH, Germany)

(Figure 3.3b). Next to an oxygen flash of 30 s that enhances the adhesion of metal to polymer, a 300-nm-thick layer of platinum is deposited via sputtering technique (Figure 3.3b) (Leybold Univex 500, Leybold GmbH, Germany). Following the lift-off process, which removes any excess metal, this step is repeated but SIROF (800 nm in thickness) is used as metal rather than platinum (Figure 3.3c). This step is optional and has been omitted in the first fabrication rounds for simplicity reasons. Again a lift-off process is necessary to remove metal residues, before the structured metal is exposed to oxygen plasma (30 s) and coated with a second layer of polyimide (Figure 3.3d). To open the electrode sites, interconnection pads and perimeters, a photo resist (AZ 9260, MicroChemicals GmbH, Ulm, Germany) is spin coated and patterned via photolithography serving as etching mask (Figure 3.3e). A reactive ion etching (RIE) process is then used to open the electrode sites and interconnection pads, and to etch the device perimeters down to the supporting wafer. At the end of the process, the single device can mechanically be peeled off the wafer (Figure 3.3f).

3.2.2 Coating of Electrode Sites

In both, recording and electrical stimulation, small platinum electrodes need to be coated to improve electrode properties with respect to low impedance and high charge injection capacity (Stieglitz 2004). Platinum black and Pt-gray were the first candidates of choice due to their known characteristics and simplicity in deposition. Both coating techniques implement an electrochemical deposition step by applying an electrical voltage on the electrodes with square-shaped voltages of different magnitudes. The materials in the electrolyte (basically platinum chloride acid) determined therefore the resulting coating. Electrolytes were synthesized for platinum gray and black according to the literature. The electrolyte for platinum gray consisted of 0.5 g platinum chloride ($PtCl_4$), 0.3 g sodium dihydrogen phosphate (NaH_2PO_4), and 6.03 g disodium hydrogen phosphate (Na_2HPO_4) dissolved in 100 ml of deionized water according to Method 1 of the US patent number 2007/0089994A1 (Zhou 2005). The electrolyte to electroplate Pt-black (Schuettler 2005) was made by dissolving 5 g hexa-chloroplatinic acid (H_2PtCl_6) in 357 ml ultrapure water and by adding 71.4 mg lead(II)-nitrate ($Pb(NO_3)_2$) (Chemicals: Merck KGaA, Darmstadt, Germany). Different TIME samples were immersed into the corresponding electrolyte. A three-electrode setup with the sample as working electrode, a large surface platinum counter electrode, and a Ag/AgCl electrode with 3 M KCl as reference were used. The working electrode was balanced with a negative square wave voltage of -0.25 V for 10 s for Pt black and -0.525 V for 30 s for Pt-gray deposition, respectively. Later in the development phase, sputtered iridium oxide (SIROF) has been used as electrode coating.

3.2.3 Electrochemical Characterization *In Vitro*

Electrochemical characterization of the electrodes has been performed to assess the transfer functions during recording and stimulation and compare the different coating processes. The electrochemical impedance of the electrodes was characterized using a gain-phase analyzer with a potentiostat (Solartron 1260 & 1287, Solartron Analytical, Farnborough, Hampshire, UK) in combination with the software ZPlot (version 2.8 by Scribner Associates Inc., Southern Pines, NC, USA) within a three electrode setup using a platinum counter electrode and a Ag/AgCl reference electrode (3 M KCl) in 0.9% NaCl solution. Impedance spectra have been taken between 10 Hz and 700 kHz applying a sine wave of 10 mV. Eight electrode sites ($\mathrm{n} = 8$) were measured per electrode material and coating. Mean and standard deviation were calculated. Lumped parameter equivalent circuit models were fitted with ZView (version 2.8 by Scribner Associates Inc., Southern Pines, NC, USA) using a simple model with the Helmholtz-capacitance in parallel to a Faraday resistance and the access resistance in series. The rationale to take an electrical rather than an electrochemical equivalent circuit was an opportunity to match electronic circuits if recording over the electrodes is intended.

Johnson's voltage noise (Liu 2008) (Equation 3.5) was calculated via the real part of the interface impedance at body temperature ($\mathrm{T} = 311$ K) with the Boltzmann's constant ($\mathrm{k} = 1.38 * 10{-}23$ J/K) in the frequency range between 300 and 1500 Hz that has been envisioned to be adequate to record either neural single unit or mass activity. The final value was calculated taking the real part of the recorded impedance at a certain frequency (Equation 3.6) and multiplying it with the related frequency "slice" ($\Delta\mathrm{f} = 1$ Hz). From these 1200 slices per electrode site, the Johnson's voltage noise was derived for a single electrode site within the defined frequency range:

$$V = \sqrt{4kT \cdot \mathrm{Re}\{\underline{Z}\} \cdot \Delta f} \tag{3.5}$$

$$\mathrm{Re}\{\underline{Z}\} = |\underline{Z}| \cdot \cos\theta \tag{3.6}$$

Noise has been calculated for platinum, platinum gray, and platinum black.

The stability of the platinum coatings, that is, the adhesion of the coating on the substrate metal, has been evaluated. Therefore, the TIME devices were placed into an ultrasonic bath (Emmi-6, 20W US, Schalltec GmbH, Walldorf, Germany) filled with deionized water. They have been exposed to ultrasound for 2, 4, and 10 min. Electrical impedance was measured before and after treatment to investigate changes in the coatings. The electrode properties during electrical stimulation have been investigated by determination of the water window during cyclic voltammetry and succeeding application of current pulses with typical envisioned stimulation parameters to determine the polarization and capacity of the electrode site. Cyclic voltammetry has been performed with the three electrode setup described above at a sweep rate of 100 mV/s. A custom made voltage controlled current source was used to apply biphasic, symmetric, rectangular, charge balanced, cathodic first current pulses with an amplitude of 100 μA and a pulse width of 200 μs, that is, 20 nC in 0.9% saline solution via block capacitors of 1 μF. A large platinum electrode was taken as counter electrode. First, electrodes have been treated with ultrasound for 2 minutes before applying current pulses. The voltage over the electrodes and the electrolyte was recorded via an oscilloscope with a sampling frequency of 1 MHz (54622D, Agilent Technologies GmbH, Böblingen, Germany). Data were transferred to a computer for offline analysis afterwards.

The platinum electrodes as well as the platinum gray and black coating showed the typical metal behavior with an access resistance of about 10 kΩ and varying cut-off frequencies and impedances (Figure 3.4, Table 3.1, mean values of $n = 8$) according to their surface roughness.

The fitting of the measurement data with the R||C-R model of the electrodes resulted in the parameters of the Helmholtz-Capacity C_H, the Faraday resistance R_F, and the access resistance R_A as shown in Table 3.2.

Thermal (Johnson) noise has been calculated in the frequency range between 300 and 1500 Hz to be (mean $\pm$ standard deviation) $V_{Pt} = 72.5 \pm 23.8$ μV for platinum, $V_{Pt\text{-}black} = 45.7 \pm 13.0$ μV for platinum black, and $V_{Pt\text{-}gray} = 52.3 \pm 15.8$ μV for platinum gray.

Stability of the electrode coatings has been investigated after ultrasound treatment for 2, 4, and 10 min by means of impedance spectroscopy (Table 3.1). The impedance remained stable for the platinum metallization. The impedance of both platinum gray and platinum black coatings increased from 22 to 33 kΩ at 1 kHz for Pt-gray (i.e. 33%) and from 15 to 22 kΩ at 1 kHz for Pt-black (i.e. 32%) after 2 min of ultrasound but stayed stable during further ultrasound treatment.

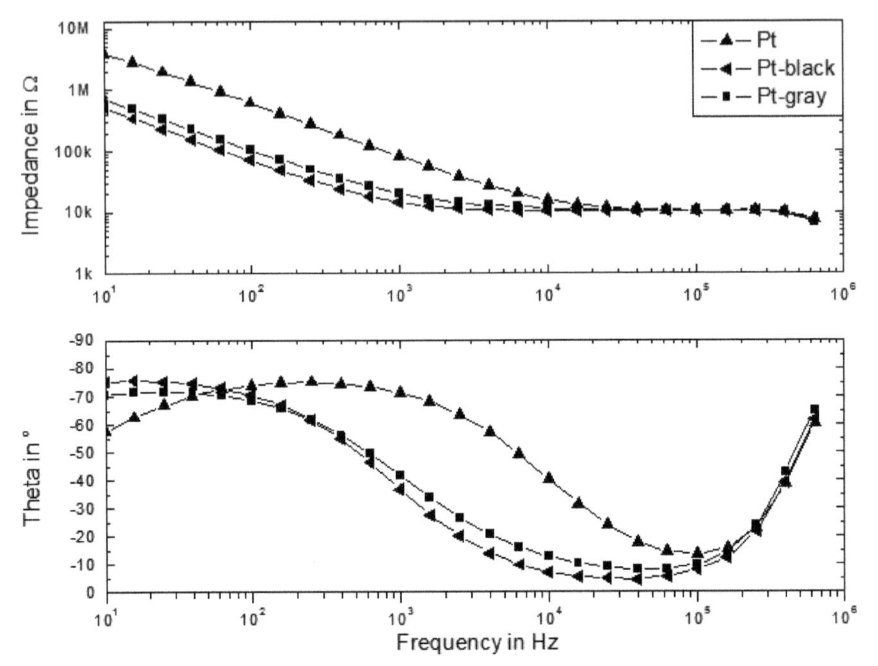

Figure 3.4 Impedance and phase angle of the TIME with platinum, platinum gray, and platinum black as electrode material.

Table 3.1 Results of EIS measurements with different electrode-coating materials

Material	$\lvert\underline{Z}\rvert$ in kΩ @ 1 kHz	Phase Angle θ in °	Cut-off Frequency in Hz
Pt	81.8	−71.4	∼10000
Pt-gray as deposited	20.2	−41.7	∼1600
Pt-gray after 2 min US	32.5	−60.8	∼2900
Pt-black as deposited	14.6	−36.7	∼1100
Pt-black after 2 min US	19.7	−50.5	∼2000

Note: Mean values of n = 8 electrodes.

Table 3.2 Lumped parameter equivalent circuit model of the electrodes fitted from impedance measurement data

Material	C_H in nF	R_F in MΩ	R_A in kΩ
Pt	6.9	1.9	10.5
Pt-gray	15.3	1.5	11.1
Pt-black	24.4	1.7	10.8

The water window for all electrodes was determined to be between 1.1 and −0.8 V with respect to Ag/AgCl. Under stimulation conditions (100 μA, 200 μs), the recorded voltage over the electrodes was analyzed (Figure 3.5).

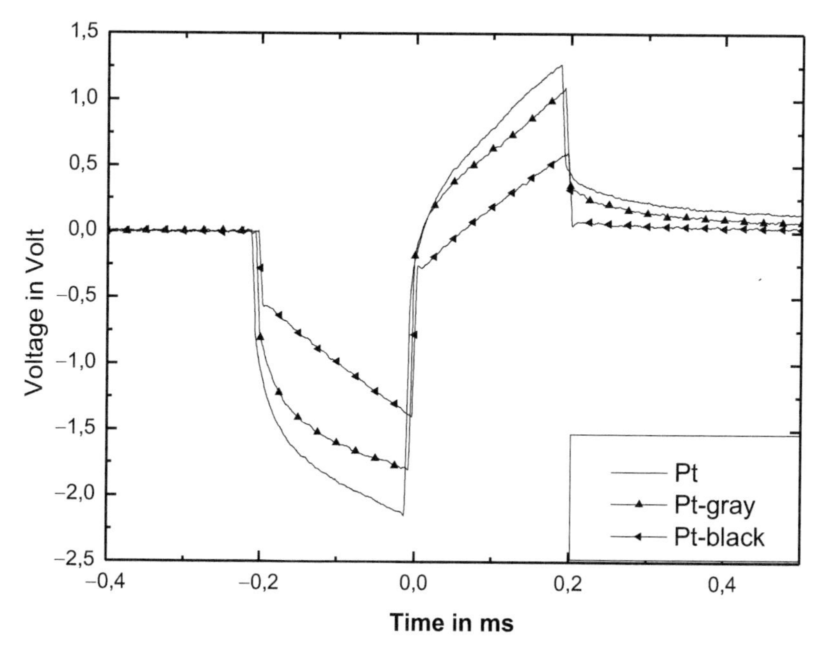

Figure 3.5 Voltage response after current stimulation of the TIME with platinum, platinum gray, and platinum black as electrode material.

Table 3.3 Voltages V_c and the derived capacitance of different materials during stimulation

Material	V_c (V)	C (nF)	$C_d(\mu F/cm^2)$	Surface Factor, n
Pt	1.56	12	424	1
Pt-gray	1.1	18	636	1.5
Pt-black	0.8	25	883	2.1

The voltage drop V_{IR} over the access resistance R_A was found to be about 560 mV in all experiments. Platinum had the largest polarization of $V_c = 1.56$ V, followed by Pt-gray ($V_c = 1.1$ V) and Pt-black ($V_c = 0.8$ V). The value for platinum has been out of the reversible region. The peak of this voltage ramp V_c, which is related to charging the Helmholtz capacitance at the phase boundary, was used to calculate the charge injection capacitance $C = Q_{inj}/V_c$ where Q_{inj} is the charge injected into the phase boundary (Table 3.3). The surface factor of the electrode coating related to pure platinum has been calculated as the charge density C_d of a given electrode divided by C_d of platinum. It indicates the enlargement factor of the electrochemical active surface of different surface qualities (Table 3.3) (Schuettler 2007, Franks 2005).

The stimulus pulse parameters in the in vitro characterization (200 μs, 100 μA) have been derived from experiences with cuff electrodes. The charge of 20 nC resulted in a charge density of about 708 $\mu C/cm^2$ of geometric electrode area. This is far beyond the electrochemical safe limits of polished platinum (Veraart 2004, Shannon 1992). The used stimulation parameters in the in vivo pilot studies (20 μs pulse width) stay within the most conservative prediction of reversible charge injection capacity of 40 $\mu C/cm^2$ (Donaldson 1986) for polished platinum if an amplitude of 500 μA is not exceeded, that is, $Q_{inj} = 35.4\ \mu C/cm^2$. Up to an amplitude of 1000 μA, that is, $Q_{inj} = 70.8\ \mu C/cm^2$, and 1500 μA, that is, $Q_{inj} = 106.2\ \mu C/cm^2$, investigations reported to be still in the electrochemical safe stimulation regime with a reversible charge injection capacity of 50–100 $\mu C/cm^2$ geometrical (Robblee 1990) and 400–500 $\mu C/cm^2$ "real" (electrochemical active) electrode area (Brummer 1977). Integration of the electrodes into a completely implantable stimulation system might afford a change of the pulse width from small pulse width that results in minimum charge of the stimulation pulse towards pulse width about the chronaxie of the stimulated nerve that results in minimum energy per stimulation pulse to maximize life time of batteries or minimize the need of inductive energy transmission. These parameters together with reduced charge carrier mobility in vivo (Cogan 2007) will lead to different reversible charge injection limits that have to be investigated in chronic studies. The increase of the electrode impedance of platinum black after explantation might indicate some instabilities in the layer. Modification of the deposition technology to ensure denser material with better adhesion (Marrese 1987) might be necessary, if platinum gray will come to its limits.

3.2.4 Assembling of Connectors and Design Optimization for First Preclinical *In Vivo* Studies

For in vitro characterization and acute in vivo experiments, a fast and easy to use interconnection is favorable. Therefore, a flip-lock zero insertion force (ZIF) connector was chosen. The assembly contains only few and simple steps (Figure 3.6). The interconnection pads are reinforced by an electrical insulation tape, inserted into the ZIF which is closed afterwards. With this technique, the whole assembly time can be reduced to just three steps, and the SMD configuration of the ZIF connectors allows multiple forms of cable connections, for example, soldering bare wires or a flat flex cable (FFC).

After first acute rat experiments, the need for a new electrode design became obvious. The TIME-1 was bulky and, when folded, rectangular

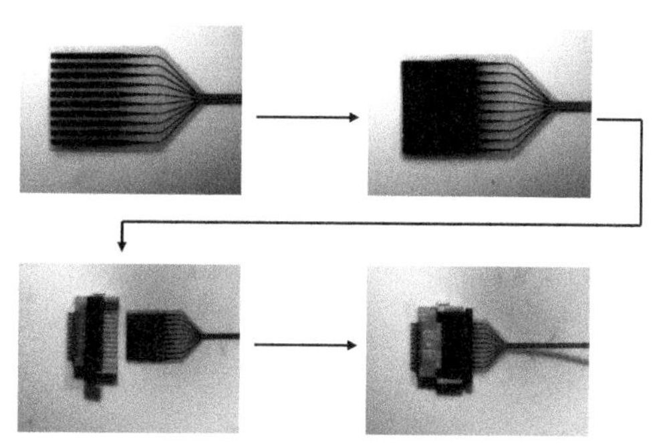

Figure 3.6 Assembling of TIME with a flip-lock ZIF connector. (a) Pad array on polyimide, (b) reinforcement with tape, (c) insertion into ZIF connector, and (d) closing of connector fixates substrate and reinforcement.

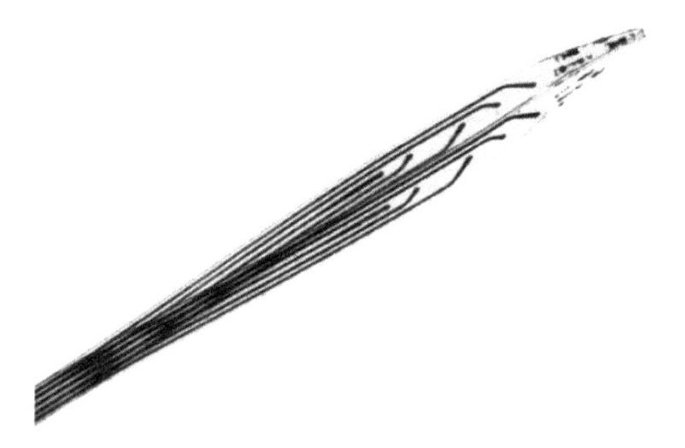

Figure 3.7 Folded TIME-1 with arrow-like tip.

in design which posed a serious problem during implantation procedure. Beneath difficulties in insertion, tremendous nerve damage was induced resulting in limping rats. Hence, a provisional solution was found in sharpening the folded electrode tip in an arrow-like structure (Figure 3.7). This improving implantation technique increased the success of implanting and stimulating the rats' sciatic nerve. Nevertheless, TIME-1 is far too wide (580 μm) for the rats' nerve and its straight arrangement of electrode sites is suboptimal. Henceforth, TIME-2 will become smaller in width, have a defined arrow shape with a tip-width of merely 100 μm, and have an increased electrode count of 3 to 6 active sites.

In general, three different versions of the TIME-2 were designed (Table 3.4): one for small animal models (rat, Figure 3.8a) and two layouts for large animal models (pig, Figure 3.9). Design specifications were derived

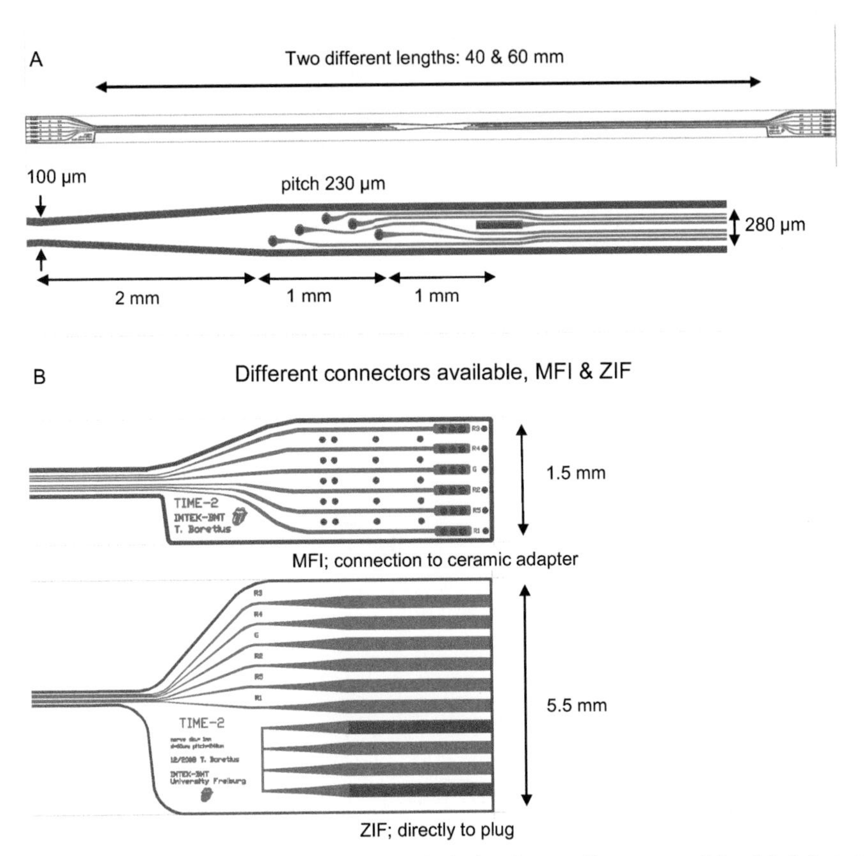

Figure 3.8 Schematic view of the TIME-2 device for small nerve models. (a) Substrate design with arrow tip and 5 electrode sites per side. (b) Different connection pad designs. Left: arrangement for MFI on ceramic substrate. Right: Arrangement for direct ZIF connection.

from anatomical studies in close cooperation with the corresponding partners (AAU and UAB). ZIF and MicroFlex Interconnection (MFI) options for assembly were realized for all versions of TIME-2 (Figure 3.8b).

TIME-1 electrodes were assembled using zero insertion force (ZIF) connectors, because of the easiness in handling and the ability to connect the devices post implantation (Figure 3.10). This technique ensures an easy implantation procedure, since the electrodes' handling is not hindered by cables and the electrodes' relative movement can be kept at a minimum. This approach is adequate for the rat animal model since the length of the integrated interconnects ensures that the cable–socket interface is outside of the animal.

Table 3.4 Specifications of TIME-2 devices

	Small Nerve	Large Nerve (Version A)	Large Nerve (Version B)
Interconnection	ZIF and MFI	MFI	MFI
Length (without interconnection)	40 and 60 mm	60 mm	60 mm
Width	280 μm	180/400 μm	260/480 μm
Width in mirror line	100 μm	100 μm	100 μm
Number of electrode sites	5	3	6
Electrode pitch	230 μm	1 mm	0.5 mm
Ø active sites	60 μm	60 μm	60 μm
GND dimensions	60 × 400 μm	0.24 × 2.4 mm	0.24 × 2.4 mm
Distance from mirror line to first electrode	2 mm	7 mm	7 mm
Distance from first to last electrode	1 mm	2 mm	2.5 mm
Track width	15 μm	15 μm	15 μm
Track pitch	15 μm	15 μm	15 μm
Referred as	TIME2-S-40/60-MFI/ZIF	TIME2-L-A	TIME2-L-B

But since the ZIF connectors are bulky and comparably heavy, a new approach was undertaken to further decrease the weight and dimension of the adapter from thin-film substrate to the actual wiring of the TIME-2 devices for a large animal model (Figure 3.11). Here, longer cables are needed to connect the electrode–cable assembly to standard plugs outside the animal. Therefore, a ceramic adapter (alumina, Al_2O_3 ceramics) was introduced which comprises Pt/Au conductive tracks and an insulating overglaze layer. This adapter is manufactured through screen printing of thick films. The technology was already established at IMTEK during the electrode assembling process and could be applied for this purpose. For the assembly of TIME-2 devices, Cooner wires (AS-632 by Cooner Wire Inc., Chatsworth, CA, USA) were soldered to the 10- or 16-way pin connector for version TIME-2 A or B, respectively. These Cooner wires were cut at a specified length of approximately 25 cm and again soldered to the two ceramic adapters (one for each side of the thin film substrate). Afterwards, the polyimide substrate was assembled onto the ceramic adapters with the MicroFlex Interconnection technique (Meyer 2001). The assembly was fixed with a tiny droplet of epoxy glue. The adapters with the attached thin films were placed back to back and fixed through a silicone tube. At last, residual openings become insulated with silicone rubber (Figure 3.11).

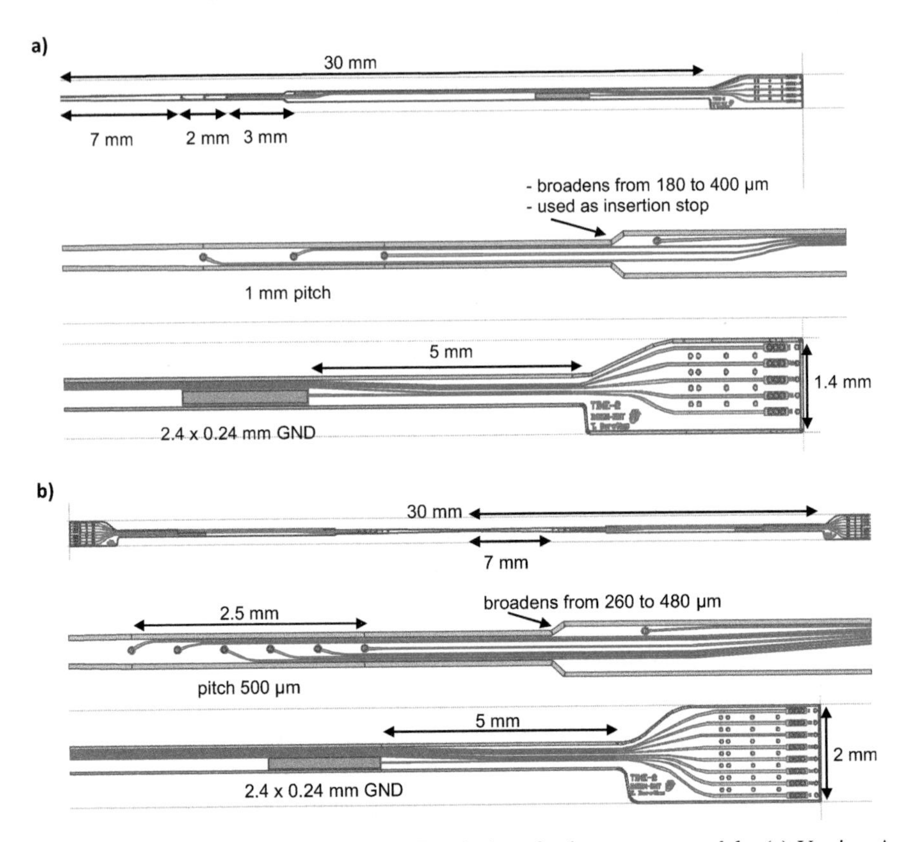

Figure 3.9 Schematic view of the TIME-2 devices for large nerve models. (a) Version A with three active sites and (b) Version B with 6 active sites.

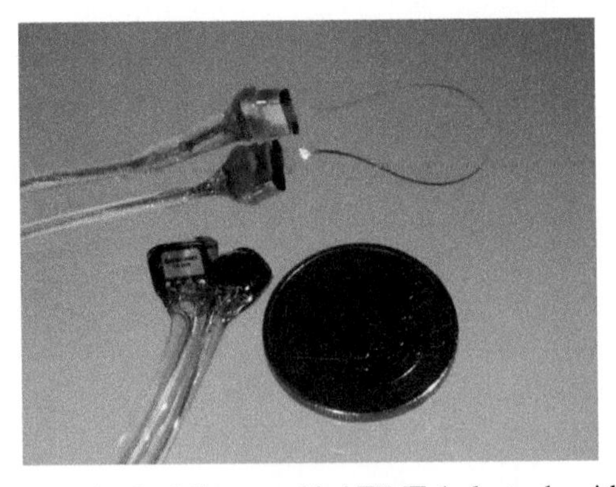

Figure 3.10 Photograph of a fully assembled TIME-1 electrode with ZIF connectors.

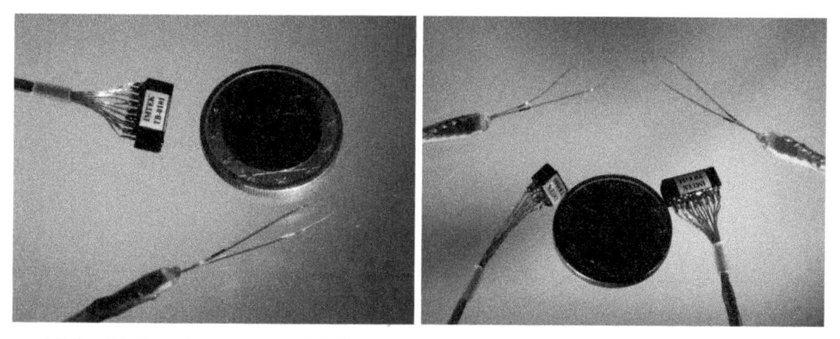

Figure 3.11 Light micrographs of fully assembled TIME-2 devices. Left: TIME-2-B; right: both versions of the TIME-2.

The devices were ready for in vitro and acute in vivo use and passed to the partners' laboratories for validation of simulation studies and acute in vivo tests (see chapters below).

One of the main challenge in manufacturing devices for chronic implantation lies within a well adherent encapsulation (Schuettler & Stieglitz 2013). Since all polymers are permeable to water molecules, the main responsibility of a non-hermetic encapsulation is to prevent condensation of water on the surface of the device. If, for example, an air bubble is enclosed within the encapsulant and the devices' surface or conducting track, water will permeate through the barrier and slowly fill the bubble. With time, the pure water will dissolve ions out of the surface of the device and, thereby, change the direction of the osmotic pressure towards the bubble itself. Consequently, more water permeates to the bubble, more ions are dissolved, and the osmotic pressure increases furthermore. Since the bubble has to expand, the encapsulant will be separated from the surface at some point, leaving an even bigger crevice behind which again fills with water. Therefore, it is mandatory to yield a clean and well adherent boundary between the devices' surface and the encapsulation. For this purpose, a cleaning and rinsing protocol was established to ensure that all residues from soldering and handling are removed prior to encapsulation.

After successful acute animal experiments, the TIME-2 was tested in chronic animal experiments. Beneath the results from stimulation and recording tests, it will show whether our protocols and precautionary measures of cleaning and encapsulation were sufficient to sustain a working device in the living body. As for the acute experiments, Cooner wires were cut and soldered to the thick-film printed ceramic adapter as before. The other ends of the

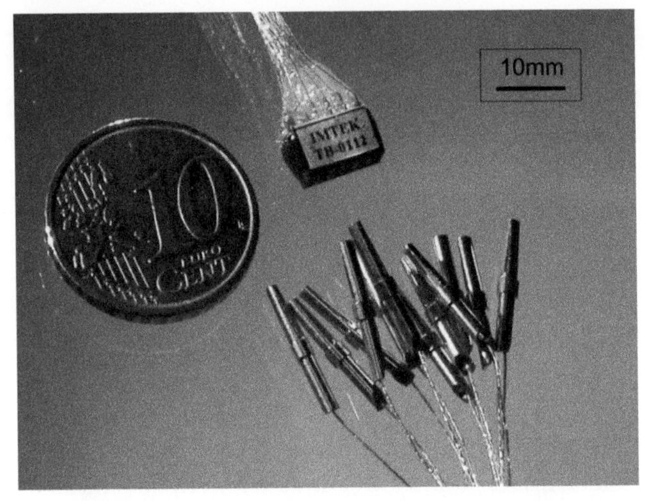

Figure 3.12 Cooner wires soldered to different plugs.

wires are soldered to two different connectors (Figure 3.12): one small, 12-way socket plug which is made out of high temperature thermoplastics and incorporates gold over nickel contacts, and 12 single pin contacts consisting also of a gold–nickel compound. The motivation to use these connectors was the need to implant the whole device including plugs, since rats quite often try to remove any percutaneously placed connector, and the ability to coat these plugs easily with silicone rubber. If a measurement would be required, the plugs could be explanted and the encapsulation removed using a scalpel. After the experiments, silicone caps can be reattached to the connector and sealed with a few droplets of silicone glue which cures within minutes. Subsequently, plugs can be buried again in tissue pockets until the next measurement is due.

When assembling of plugs and ceramic adapters was finished, the complete structure had to be thoroughly cleaned. To ensure an overall immaculate surface, all steps afterwards are conducted in a class 10000 clean room. The steps of the cleaning process included the following:

- Removal of flux residues from soldering with flux-remover (de-flux 160 by Servisol, Iffezheim, Germany)
- Removal of flux-remover residues by washing with isopropyl alcohol (IPA)
- Removal of IPA residues by washing in deionized water

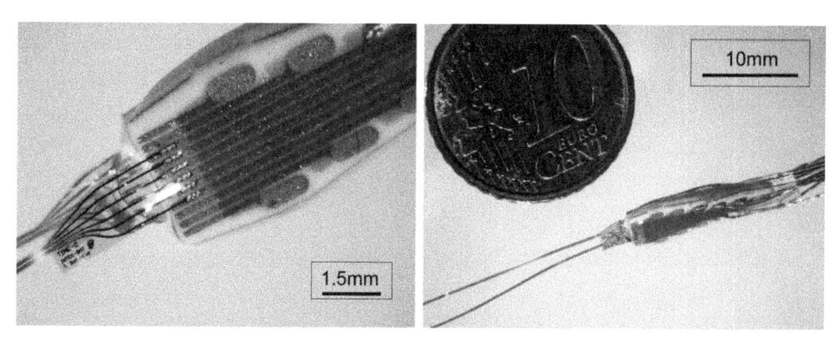

Figure 3.13 Completely encapsulated TIME-2 electrode. Left: MicroFlex Interconnections between thin-film and ceramics. Right: Overview of TIME, adapter, and wires.

- Removal of any other residues by slewing device five minutes in Leslie's Soup (Leslie's Soup is named after a British researcher and consists of the glass cleaner "Teepol-L" (0.5%), Na_3PO_4 (2.5%), and deionized water (97%)).
- Removal of residues by rinsing with deionized water
- Washing in deionized water till the conductance value of DI water settles again
- Drying on lint-free cloth at 60°C in an oven for at least 3 hours

After transferring the cleaned structures into the clean room, the actual thin-film electrodes were connected via MFI technique to the ceramic adapters with the soldered wires. The two separated adapters are placed back to back and glued via a droplet of silicone rubber. Following the curing period which hardens the polymer, the ceramic adapters are encapsulated in medical grade silicone (MED-1000 by PolyTec PT GmbH, Waldbronn, Germany).

Special attention is hereby dedicated to ensure a void-free film by applying first a more viscous solution of silicone (50% silicone and 50% n-heptane) to encapsulate the small structures around the solder joints and MicroFlex Interconnections. A curing step is followed by a second layer of encapsulation (100% silicone) to encompass the complete adapter (Figure 3.13). The plugs are processed in a similar manner later on, when the adapters are fully cured.

3.3 From Flat to Corrugated Intrafascicular Electrodes

In addition to the TIME's ability to access deeper nerve fibers in the nerve trunk, the corrugation of the TIME structure is expected to not only lead to better selectivity and electrode–tissue contact but also stabilize the electrode structure by preventing device movement after implantation. Preliminary studies on the corrugation of electrodes were conducted by Bossi et al. (2007, 2009). Shape memory alloys (SMA) were added to the tf-LIFEs to add micro-actuation in these studies. Concerning the corrugation dimensions, they achieved amplitudes of around 10 μm (Bossi et al. 2007) and 20 μm (Bossi et al. 2009), where half-wavelength was around 500 μm in both cases. These corrugated LIFE structures, however, were not tested in vivo.

3.3.1 Design Considerations

The design guidelines and desired target specifications for corrugated TIMEs included: (1) low power consumption during the implantation period with respect to corrugation and actuation; (2) temperature increase should be kept as low as possible considering thermal damage during the implantation period when active corrugation is used; and (3) simple structure for easy fabrication and good biocompatibility and long-term mechanical stability. Among the well-known micro-actuation principles (Kovacs 1998), electrostatic and electromagnetic actuation principles are not suitable in our case because of their high actuation voltage (or current). Thermal actuation principles, including thermal bimorph and SMA, could possibly cause heat damage to surrounding tissue. Piezoelectric actuation could be a good candidate because it has a medium-range actuation voltage and no thermal issues. The multi-layer structure of these actuators, however, together with the fact that they are made from novel materials may cause long-term stability and toxicology problems. Moreover, in all cases, the issue of power consumption cannot be avoided and integration of actuators significantly changes the mechanical properties of the implantable devices and thereby their structural biocompatibility. In this context, we chose the 'shaping of flat electrode through pre-actuation' strategy instead of integrating actuators in the electrodes. For the pre-shaping of electrodes, we tried three approaches: a precision machining approach, a micromachining approach, and a hybrid of precision and micromachining approach. The details and results of these approaches will be described in the following sections. The pre-shaped electrodes are to be implanted in their original corrugated shape. Therefore, no additional power consumption is needed after implantation for corrugation. Thanks to the inherent flexibility of the thin film polymer substrate, the pre-shaped electrode substrate can be stretched to 'mild-corrugated' shape during the insertion period and then return to its corrugated shape after insertion, minimizing cell damage during the insertion period. The structure and materials used in these approaches

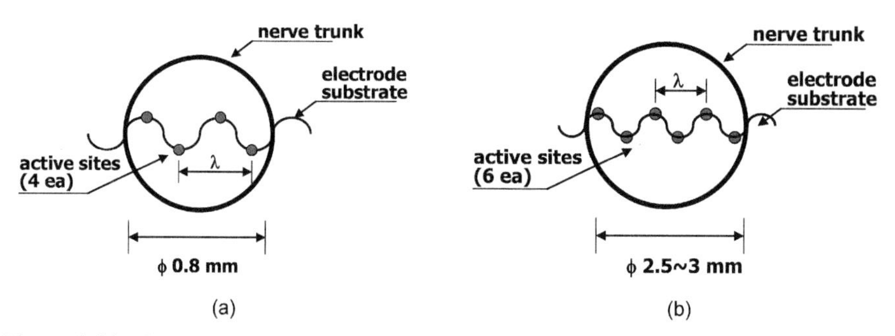

Figure 3.14 Schematic view of the cross section of nerve trunk with corrugated electrode: (a) in small animal model, which has 4 active sites; (b) in large animal mode, which has 6 active sites.

Table 3.5 Comparison of required corrugation dimensions and achieved ones by various manufacturing technologies

	Required for		Achieved by			
Parameters	Small Animal	Large Animal	SMA (Bossi et al. 2007, 2009)	Precision Mechanics	Micro-machining	Hybrid Machining
Wavelength (λ)	$\approx$400 μm	850 μm ~ 1000 μm	$\approx$1 mm	500 μm ~ 4000 μm	300 μm ~ 400 μm	400 μm ~ 1500 μm
Amplitude (h)	>50 μm	>100 μm	10 μm ~ 20 μm	90 μm ~ 700 μm	$\approx$20 μm	>50 μm

remain mainly the same as within the basic TIME, which means no additional mechanical influence and cytotoxicity concerns. One possible drawback of this approach is passivity of corrugation parameters (dimension of waveform). The dimension of corrugation, however, can be tested and optimized during the animal test phase, and the results will be directly reflected in the fabrication process.

To derive corrugation parameters (wavelength: λ, and amplitude: h), we assumed four active sites in the small animal model (rat) and six active sites in the large animal model (pig). Figure 3.14 shows a schematic cross-sectional view of a nerve trunk with a corrugated TIME. Each red dot indicates an electrode's active site. The nerve diameter for the small animal model was assumed to be 0.8 mm and that of the large animal was assumed to be 2.5–3.0 mm based on anatomical studies. Considering these parameters, the required wavelength (λ) of corrugation was calculated as 400 μm for the small animal model and 850–1000 μm for the large animal model. In terms of corrugation amplitude (h), we decided to fix the minimum requirement as 50 μm in the small animal model and 100 μm in the large animal model, according to our experiences (see Table 3.5). The detailed design specification had to be determined from the chronic test phase.

It is noteworthy that we devoted ourselves to developing manufacturing technologies, which can cover a wide range of dimensional spectra. While we did not yet have the specific dimensional requirements before the animal experiments started, some of the technical approaches developed had to be adoptable to the final specific needs.

3.3.2 Precision Machining Approach

Similar to the thin-film cuff electrode manufacture, this approach can be achieved through a temper process that comprises a moldform and the already fabricated flat TIME substrate. Since the moldform is solely fabricated via precision mechanics, variable dimensions can easily be obtained, yielding in low cost and rapid prototyping (Boretius 2009).

Three moldforms were designed and fabricated through wire-cut electrical discharge machining (EDM). The forms comprise a medical grade steel mold (1.4541 by Goodfellow GmbH, Bad Nauheim, Germany) and an aluminum housing.

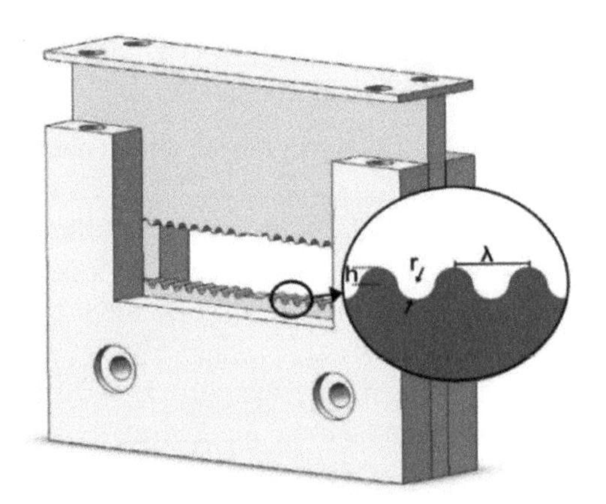

Figure 3.15 Schematic view of the moldform's design.

Source: Reprinted with permission from Boretius, T., Zimmermann, D., and Stieglitz, T. Development of a corrugated polyimide-based electrode for intrafascicular use in peripheral nerves. IFMBE Proceedings 25/IX, Springer, Munich, Germany, pp. 32–35 (2009).

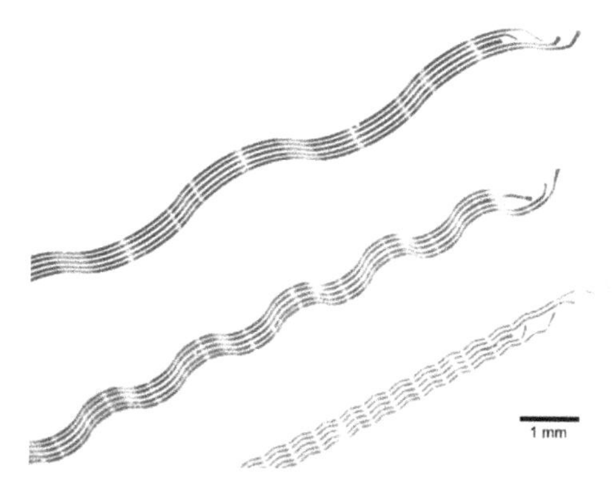

Figure 3.16 Prototypes of corrugated electrodes using precision machining approach.

The three forms of different dimensions were fabricated in this first trial. All comprise a sinusoidal form defined by a wavelength λ and an amplitude h (Figure 3.15). The manufacturing of the test structure was performed using standard TIME manufacturing processes (see above and Boretius 2010). The test structures were mechanically peeled off the wafer using tweezers and inserted into the mold. The mold was closed afterwards and fixed via screws. The complete form was transferred into an oven (PEO-601, ATV Technologie GmbH, Vaterstetten, Germany) and tempered in an inert nitrogen atmosphere at 300°C for 90 minutes. After the cooling period, the mold was opened again and the corrugated electrode was retrieved.

Electrodes have been fabricated and successfully corrugated (Figure 3.16). Contour accuracy and strength of shape of corrugated devices were obtained optically and compared to the original moldforms' dimensions. Amplitude and wavelength were separately measured for tempered electrodes, yielding the contour accuracy, and for devices stored for 2 weeks in saline solution at 37, 60, and 85°C, which gives the strength of shape. Regarding the amplitude, between ~91% and ~51% of the moldforms' dimension could be transferred into the polyimide substrate after the temper process, whereas the strength of shape decreased (~3.3% to ~24.8%) with the storage at different temperatures. The transferred wavelength was more accurate with ~98% to 90% for the small and large molds. In this case, only 0.3% to 9.4% strength of shape was lost while stored at 85°C for 2 weeks. After the temper process, 80% of the amplitude and 98.5% of the wavelength were transferred into the substrate. The storage time decreased the wavelength of about 4.6% and the amplitude of about 13.4% further (Boretius 2009).

It is important to note that corrugation within the small mold, which was designed for small animal usage, yielded a rather trapezoidal structure instead of a sinusoidal one. Since this trapezoidal form has an extreme high strength of shape, it is nearly impossible to "decorrugate" the substrate during implantation and handling. Hence, it can be concluded that substantial nerve damage would be introduced using this small mold to corrugate TIME electrodes. Larger molds yielded smoother corrugations, and the sinusoidal form was also implemented within the substrates. Although the dimensions of corrugation would fit large nerve models, this approach was neglected because of the observation of microcracks within the platinum layer due to mechanical stress (thermal expansion) of moldform and polyimide substrate. To overcome these limitations, a comb like mold was designed in which the polyimide substrate comes to rest in mid-air, and thus, no mechanical stress can take effect (see Section 4.3 in Chapter 4).

3.3.3 Micromachining Approach

In the previous precision mechanics approach, the achieved amplitude values were satisfactory. The smallest wavelength, however, which is limited by the precision of EDM process, was marginal, especially for small animal application. For the smaller wavelength, we have applied a thin-film micromachining process. Here the dimensions of corrugation are defined by lithography technique. Thus, the smallest feature size can be downsized dramatically compared to precision mechanics. General thin-film micromachining technique, however, has difficulty in achieving out-of-plane three-dimensional structure, like corrugated electrodes. To tackle this problem, we used deformation induced by unevenly distributed residual stress.

Residual stress layers were incorporated into the flexible substrate of the TIMEs (Figure 3.17). In contrast to the basic TIME, high residual stress layers (red color in Figure 3.17a) are placed on top and bottom side of electrodes. Regardless of compressive or tensile residual stress, the residual stress layer induces bending of substrate layer under (or over) it, because the position of residual stress layer is far from the geometrical center (i.e., the neutral fiber) of the substrate layer. We simplified the structure (Figure 3.17b) without difficulty making the substrate instead of the stress layers asymmetric. The resulting radius of curvature, R, can be derived from the beam bending theory as:

$$R = \frac{E_s t_s^2}{6 t_f \sigma_f} \tag{3.7}$$

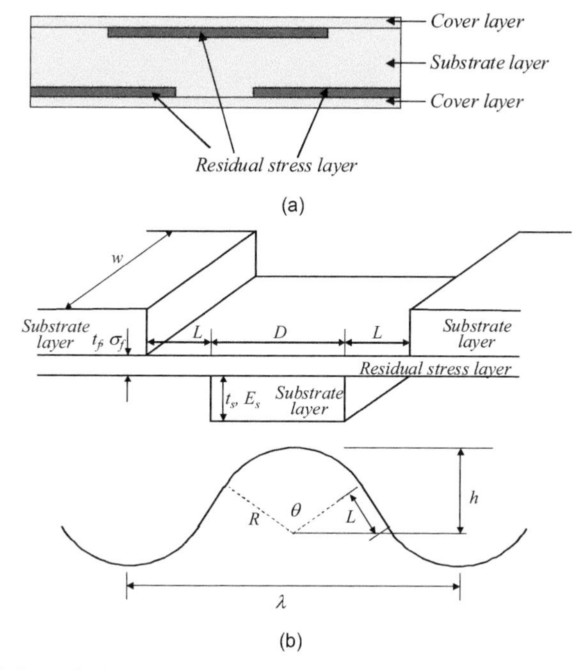

Figure 3.17 Schematic drawing of electrode to be corrugated using residual stress. (a) Cross-sectional view of electrode structure. High residual stress layer (red layer) is placed on both side; (b) Simplified model of structure in (a) (top) and its deformed shape (bottom). Residual stress layer is placed in the middle and the substrate layer is placed on either top side or bottom side of the residual stress layer repeatedly.

where E_s denotes Young's modulus of substrate, t_s and t_f are thickness of substrate and residual stress layer, respectively, and σ_f is the residual stress in the residual stress layer. From the geometrical relation (Figure 3.17b, bottom), the wavelength, λ, and amplitude, h, can be calculated as:

$$\begin{aligned} \theta &= D/R \\ \lambda &= 2 \times (2R\sin\theta/2 + L\cos\theta/2) \\ h &= (R - R\cos\theta/2) + L/2\sin\theta/2 \end{aligned} \tag{3.8}$$

where D and L are denoted in Figure 3.16b.

Table 3.6 Design parameters and expected results of the corrugation induced by residual stress

	Design Parameters (Denoted in Figure 3.17b)		Expected Results	
Device Type	*D*	*L*	Wavelength, λ	Amplitude, *h*
Type 1	150 μm	150 μm	≈400 μm	≈100 μm
Type 2	150 μm	75 μm	≈325 μm	≈68 μm
Type 3	200 μm	200 μm	≈355 μm	≈157 μm
Type 4	200 μm	100 μm	≈319 μm	≈108 μm

We designed four different kinds of test structure depending on various D and L values. The designed dimensions and expected wavelength and amplitude are listed in Table 3.6. The residual stress was assumed as 1 GPa, and the other parameters, for example, thickness of substrate layer, came from basic TIME design.

Silicon nitride (SiNx) was the material of choice as a high residual stress layer. Silicon nitride can tolerate high residual stresses in thin layers. It is also known that silicon nitride is a biocompatible material (Neumann 2004; Kue 1999). However, potential materials for a residual stress layer are not limited to SiNx. If it has high residual stress and its fabrication process is compatible with that of basic TIME, any material such as metal or polymer can also be applied. The manufacturing process was carried out in a class 1000 clean room. The manufacturing process (Figure 3.18) started with spin coating of a thin polyimide layer (<2 μm) (Figure 3.18a). The first silicon nitride layer was deposited using plasma-enhanced chemical vapor deposition (PECVD) and patterned photolithographically (Figure 3.18b). The structural polyimide layer (≈3 μm) was coated (Figure 3.18c) and followed by second silicon nitride deposition and patterning (Figure 3.18d). Finally, a polyimide covering layer was coated (Figure 3.18e), and the perimeter was etched via reactive ion etching (RIE).

After the release from the wafer, the test devices were deformed and corrugated as expected (Figure 3.19). Corrugation parameters were measured to be smaller than the estimated values. Especially the amplitude seemed not exceeding several tens of micrometers. The reason for this discrepancy could be due to the fact that: (1) the residual stress in silicon nitride was not as high as expected; (2) the residual stress in silicon nitride was relaxed due to the high process temperature (350°C) in the following polyimide curing process; (3) residual stress in the silicon nitride layer was compensated by opposite residual stress in polyimide layer.

More detailed modeling and measurements would be able to better understand shape fidelity, stress development, and release.

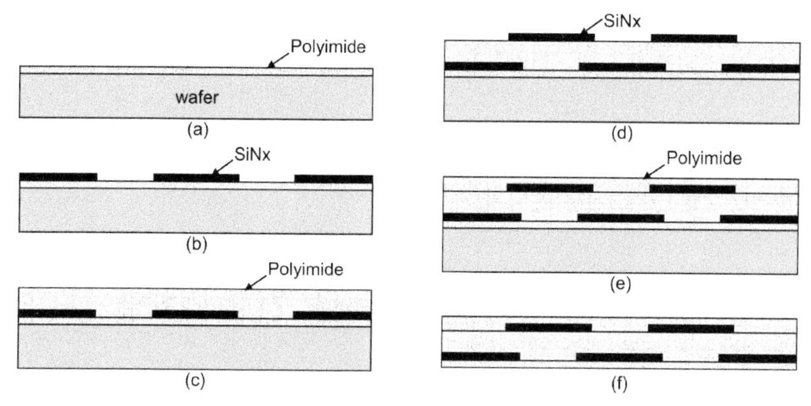

Figure 3.18 Thin-film micromachining process for corrugated electrode.

Figure 3.19 Photograph of the corrugation test structure fabricated by micromachining technique.

3.3.4 Precision Mechanics and Micromachining Hybrid Approach

The third approach was a trade-off between the first two approaches. In the precision mechanics approach, high aspect ratio (large amplitude) was possible compared to micromachining. The wavelength, however, was limited by the EDM process compared to the micromachining process. In the hybrid approach, we were pursuing fabrication of the moldform using a micromachining technique, and the corrugation of the electrode would have been achieved by temper processing like in precision mechanics. The moldform is to be fabricated by photolithographical patterning of thick-film photoresist (SU-8) and copper (Cu) electroplating, which should overcome the limits of precision mechanics in defining corrugation patterns. Furthermore, a comb-like structure will be implemented to resolve mechanical stress issues as seen in the closed moldform version. Here, the polyimide substrate is inserted into the comb mold and gets fixed at the outer perimeters (ribbon and bight), while the actual part to be corrugated rests in mid-air within the comb. Thus, all designated amplitudes and wavelengths should be feasible without the impact of mechanical stresses or additional manufacturing steps.

3.3.5 Final Decision on Corrugation Processes for Various Nerve Diameters

Three different approaches to corrugate flexible electrode substrates were developed and evaluated in total (Table 3.5). In the first approach, a moldform was made by precision mechanics and the electrode was corrugated using temper process in the moldform. This approach worked for 'large-scale' corrugation, that is, wavelengths of about 1 mm (minimum value was 500 μm) with different possible amplitudes. This technique, however, did not meet the requirements for dimensions needed within TIME electrodes for small animal models. Second, a micromachining approach was utilized which uses different residual film stresses to corrugate the electrode. A silicon nitride (SiNx) polyimide stack was produced with defined dimensions to yield the desired amplitude and wavelength of corrugation. The achieved wavelengths were about 300 μm, which would fit the small nerve model. The amplitudes, however, were limited as merely tenths of microns, which was found to be too small, since fascicle distributions within nerves are to be expected in the hundreds of microns. Third, a hybrid approach of precision and micromachining has been designed using an electroplated comb-like moldform by patterning a SU-8 layer, followed by copper electroplating. This technique is expected to overcome the limits of precision mechanics in defining the mold and corrugation pattern and, thus, should yield in a trade-off between the first two corrugation techniques.

All techniques described above are developed to cover different ranges of corrugation dimensions (wavelength and amplitude) but no one would fit all nerve dimensions. The micromachining approach is intended for the small animal model and the other two for the large animal model. Each of them showed feasibility to each application area. However, we found many technical difficulties in developing corrugated electrode for small animal model. Developing corrugated electrodes for small nerve models would demand completely different approaches that would be cost and time-consuming and would not be transferrable to larger nerves. It is important to mention that all techniques described above are feasible for use within large nerve models in preclinical trials in large animals and clinical trials in humans which is in accordance with the ultimate goal of the TIME project. Results from simulation experiments indicated no significant difference in stimulation threshold decrease when using corrugated electrodes in comparison to flat ones. Since the goal of the project was to investigate a treatment of phantom limb pain and not a comprehensive design study on micromachining techniques, we eventually decided not to spend more time on the development of corrugation strategies but on a flat design and its assembly with cables to be able to stimulate large nerves on a fascicular level.

3.4 From First Prototypes to Chronically Implantable Devices

When manufacturing chronic implants, many different considerations have to be taken into account like the overall design, encapsulation, choice of materials, cleaning protocols, and handling. While many of these aspects were already considered in building up chronic electrodes for small animal models (Section 4.2), great efforts were invested to develop alternative connectors and cables for a large animal model. Suitable connectors should be placed protectively under the skin so the animal can move freely and is not able to root or bite the electrode system out by itself. However, the connector must also be readily accessible for recording and stimulation experiments without the need of time-consuming surgeries. Therefore, a percutaneously placed connector was chosen for chronic pig experiments, which exits the skin in the neck and is fixed to the skin via a custom build metal housing that is sutured in place. This concept of percutaneous cables is identical to the envisioned human implantation in which the cable passes the skin and the connector is outside the body. On the same end, cable and wires have to be built in a way so they can give the possible highest degree of flexibility and stretchability. Since the cable is rooted through the animal and will mediate any movement to the electrodes' and connectors' placement, it is mandatory to include means of stress and strain relief. Henceforth, the developmental outcome will be depicted in detail within the next sections. Additionally, the TIME-3 electrode is introduced, and the improvements in comparison to TIME-2 devices are explained.

Because chronic implants in rats did not show reactions with respect to residues after cleansing (i.e., they have been cleaned in a proper and adequate manner) or degrading/detaching encapsulations, the protocols for cleaning and rinsing during manufacture and encapsulation techniques were maintained as described above. Additionally, the complete assembled electrode is purged with alcohol (70% ethanol) within a cleanroom environment and disinfected within an implant washing machine (Miele, Germany) as a last step before it is dried again and double-bagged in sterilizable pouches. Steam sterilization itself is carried out at the institutions of the partners performing large animal experiments.

3.4.1 Design Changes Towards TIME-3

After the initial acute experiments in small and large animal models (rats and pigs) as well as in chronic implantations (rats), first drawbacks became obvious within the TIME-2 design. Since the polyimide ribbon was completely straight in design with cables orthogonal to the peripheral nerve, the implanted device had degrees of freedom in all three directions and, thus, could move in relation to the surrounding tissue, which implies the danger of deranging the location of the active sites and/or inducing nerve damage. Additionally, TIME-2 electrodes incorporated no structures for fixation, which promoted the aforementioned problem even further. Although stimulation could be successfully administered within these initial animal models, the maximal capacity of injectable charge was restricted to small values due to the use of platinum. Values of charge injection capacities of platinum were measured to approximately 75 $\mu C/cm^2$, which is in accordance with the values found in literature (Craggs 1986). With respect to the area of one active site (Ø80 μm), this yields in only 4 nC of injectable charge. According to human trials with thin-film LIFE electrodes (Rossini 2010), this is definitely insufficient for the stimulation of human nerves over a period of 4 weeks. Dhillon and Horch reported that values of 5 nC only after 2 weeks were necessary to elicit sensations in human amputees and values up to 60 nC were used in trials to detect the limits of non-painful stimulation with their LIFE electrodes (Dhillon 2005).

To overcome these problems, a new design of TIME-3 electrodes was developed in collaboration with all partners participating in preclinical and

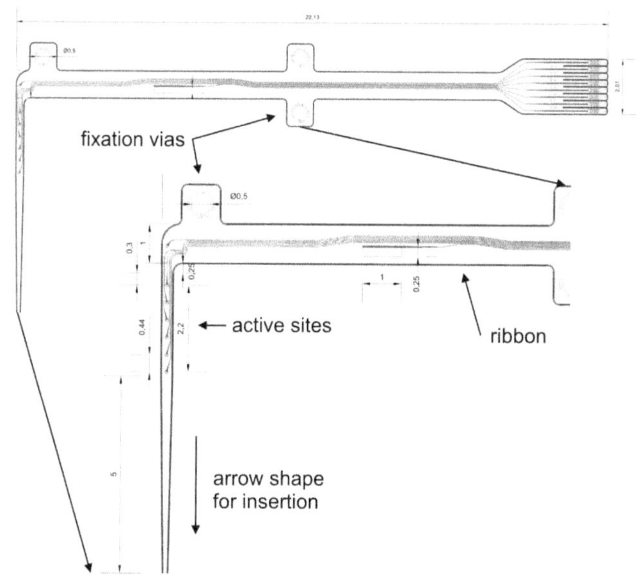

Figure 3.20 Schematic side view of a TIME-3 electrode. Fixation flaps have been integrated; 90° angle between intraneural linear array and interconnection lines reduces movement opportunities after implantation. Units in mm.

clinical trials. To minimize movement and improve fixation, a 90° bend was introduced to the electrode that separates the ribbon and the actual implanted lip (area of active sites and arrow shape; see Figure 3.20). For fixation, three vias (i.e. holes) were incorporated into the device. To increase the visibility of these vias during implantation, a shiny platinum ring encompasses the fixation structure. With these modifications, the electrode can now be implanted in a true transversal way and the ribbon can be routed distally from the implantation site in parallel to the nerve surface where it can be fixed to the surrounding tissue (epineurium). For strain relief, the cable is routed a few centimeters more in the distal direction before turning 180° to the targeted area of connector placement (in the neck of the animal). To increase the maximal charge injection capacity, sputtered iridium oxide films (SIROF) were incorporated as active site metallization. SIROF benefits from its high intrinsic porosity that increases the electrochemical active surface area (not the geometrical area) and faradaic charge transfer mechanisms (Cogan 2008). Therefore, a superior charge injection capacity is achieved in comparison to platinum. The values were again measured to approximately 2.3 $\mathrm{mC/cm^2}$, which yields in more than 100 nC of injectable charge for an electrode with a diameter of 80 μm. So even with common safety margins (75% of maximal charge injection), sufficient charge capacities should be available for chronic pig experiments and human trials.

3.4.2 Development of Helical Multistrand Cables

Within the very first acute studies in rats, insulated copper wires were used as cable strands because of their easy handling and soldering (see Figure 3.21 top). But due to the fact that copper wires are toxic in the case of a first failure and tend to corrode, there is no mean to use them within implants that should last days or even months in vivo. Therefore, Cooner wires were used as standard wires in all the following experiments (see Figure 3.21 bottom). Their properties are well known, since they are also commonly used as leads within pacemakers for years. Cooner wires consist of 10–40 stainless steel strands (15 within AS632 wires) that are braided into a single wire and are insulated with fluorinated ethylene propylene (FEP). The overall diameter of one such wire is approximately 120 μm. Although they meet all the requirements mandatory for chronic implanted wires, there are still

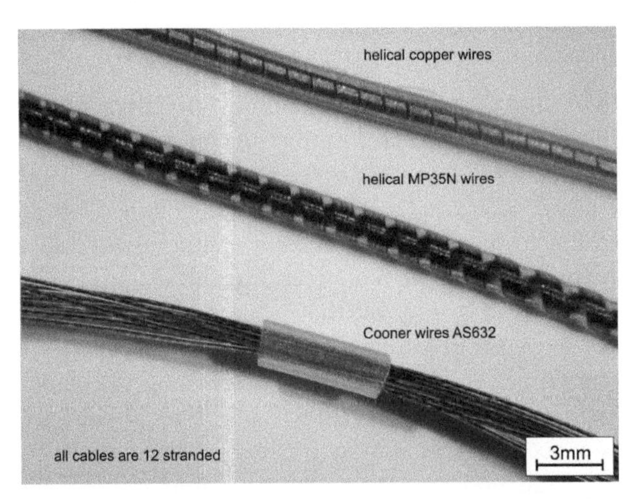

Figure 3.21 Comparison of different cable assemblies. Top: 12 helical copper wires in silicone tube; center: 12 helical MP35N wires in silicone tube; and bottom: 12 bundled Cooner wires type AS632.

Source: Reprinted with permission from Stieglitz, T., Schuettler, M., Rubehn, B., Boretius, T., Badia, J., Navarro, X. Evaluation of polyimide as substrate material for electrodes to interface the peripheral nervous system. Proceedings of the IEEE-EMBS Conference on Neural Engineering, April 27–May 1, 2011, Cancun, Mexico, 2011, pp. 529–533. Copyright 2011 IEEE.

drawbacks to be considered. First, 120 μm in diameter is still bulky when more than about 6 wires have to be used. Second, when placed into an additional silicone tube for further protection (and to avoid the incorporation of tissue into the wire bundle), the cable gets quite stiff which prevents easy routing throughout the body. And third, given that the wires are already braided, there is no possibility to incorporate structures for strain relief, which is mandatory in peripheral nerve implants.

Hence, a third alternative was necessary that encompasses all properties and requirements. MP35N was therefore the material of choice, because of its comparable properties to Cooner wires and its established use in clinical applications like the electrode cables of the cardiac pacemaker. MP35N is a nickel–cobalt base alloy that has a unique combination of properties like ultra-high strength, toughness, ductility, and outstanding corrosion resistance. One strand is insulated with polyesterimide (PEI) and has an overall diameter of merely 75 μm (Polyfil AG, Zug, Switzerland), which makes it suitable for multi-stranded wires. To add strain relief to these wires, a custom build machine was acquired (Figure 3.22 left) that is able to helically wind up to 18 single strands in parallel with a defined pitch between the single packages of strands (see Figure 3.21 center). Afterwards, a swelled silicone tube (swelling agent is n-heptane) is slipped over the helix and arbor and is allowed to shrink again to its original size. Since the arbors' thickness (plus wire diameters) is specifically chosen to match the inner diameter of the silicone tube, a tight and uniform helix is created and the arbor can be removed safely (Figure 3.22 right). This spring-like structure in the silicone tube makes the cable handy and easily routable. Furthermore, the silicone tubing is commercially available in medical grades and can incorporate X-ray contrast strips which makes the cable more visible in follow-up examinations as well as plates with serial

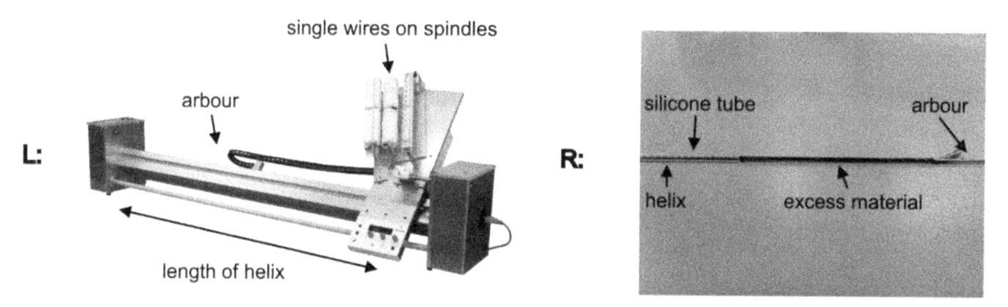

Figure 3.22 Left: Custom build cable winder with attached arbor and 16 single wires on spindles. Right: Arbor with helix of 16 MP35N wires and silicone tubing. Excess material is cut after the arbor is removed.

numbers. The feature of strain relief was investigated per tensile test, and the results proved that the helical cable could be elongated to about 150% of its initial length without any loss of electrical contact. Further tests according to pacemaker standards were successfully carried out to test the cables flexibility to bending and lifetime (see Section 4.5 on lifetime testing).

The wires have to be attached to the actual thin-film electrode. Since direct soldering of wire to thin film is not reliable, a ceramic adapter is used to mediate the connection. These adapters are manufactured in two steps: (1) a patterned layer of conductive PtAu paste is screen printed onto a small ceramic plate (2 inch) and fired within an oven at 1000°C, and (2) to insulate the single tracks, a patterned overglaze layer of glass composite is screen printed onto the ceramic and first layer. This overglaze layer insulates the complete adapter, despite specific solder and microflex spots, to which the wires and electrode are attached in a later step. After firing the overglaze layer at 750°C, the ceramic adapters are separated from the plate by a laser and are ready to use. Attachment of MP35N wires to the ceramic is done by soldering using a SnAgCu solder, because of its lead-free formula.

3.4.3 Connector Development

Within the initial studies, a classical epoxy-based connector was used because of its simplicity to use, low cost, and sufficient stability in acute experiments (see Figure 3.23 left, upper row). In chronic implants, however, the complete cable including connector has to be implanted subcutaneously (into a tissue pocket) to provide the animal with an adequate freedom to move. Additionally, animals tend to gnaw and/or shrub at transcutaneously placed connectors until this "foreign object" comes off, which will dislocate or even destroy the implanted device. Therefore, a new kind of connector was developed that uses simple stainless steel metal rings as electrical contacts (see Figure 3.23 left, lower row and right). These metal rings are placed onto a silicone tubing through a reducer chipper and are contacted via wire welding. The main properties of the rings are corrosion resistance, easy cleaning (body fluids, remains of tissue, etc.), and MRI compatibility. For further reinforcement of the connectors, the silicone tube is filled up with silicone rubber in region of the metal rings. Electrical contact is achieved through a custom build guiding structure made out of PMMA in which the metal ring contacts are inserted (see Figure 3.23 right). This guiding structure including cable is then plugged into an epoxy-based connector that incorporates protruding pins,

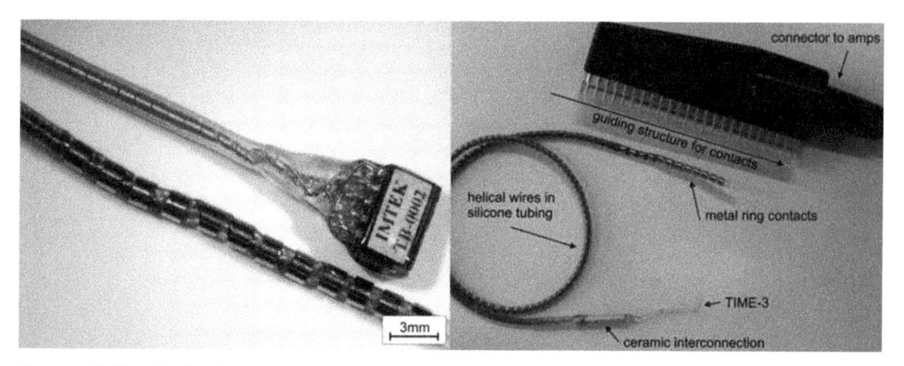

Figure 3.23 Left: Comparison between classic epoxy-based connector and metal rings. Right: Completely assembled TIME-3 electrode for chronic implants including a connector.

which ultimately establish the electrical connection with the metal rings. This system is easy to handle, is easy to clean after experiments (rinse with water or light solvents), and can be adapted to all kinds of pole configurations.

In the first chronic animal experiments with pigs, a lot of connectors with the metal rings have gone lost due to scrubbing of the animal on the ground and to the wall. Therefore, miniaturization and protection of the connector were necessary. Rings were replaced by smaller connectors which are commercially available and standard in neuroscience but unfortunately not available in medical grade quality.

The chosen connector is a commercially available Omnetics nanoplastic circular (NPC) connector with 16 poles (Figure 3.24, left). Its dimensions are 4.6 mm in diameter with a length of 5.6 mm. Cables are easily attached to the connector by soldering, and protection is provided via encapsulation with silicone rubber. Furthermore, a tag with the specific serial number of the electrode is implemented within the silicone tubing just before the Omnetics connector (Figures 3.24 and 3.25). The counterpart is also an Omnetics nanoplastic circular connector, but the male version, which has protruding pins that make contact to the electrodes' connector. A "nose" within both connectors defines the angle of insertion and ascertains correct alignment. The external cable, which is soldered to this counterpart and is running to the stimulator, can be terminated with all kinds of connectors by soldering or welding techniques. Depending on the stimulator in use, this external cable can easily be produced and distributed.

In addition, a custom build metal housing was developed by the partners who performed the pig experiments. It gets sutured to the skin and can hold

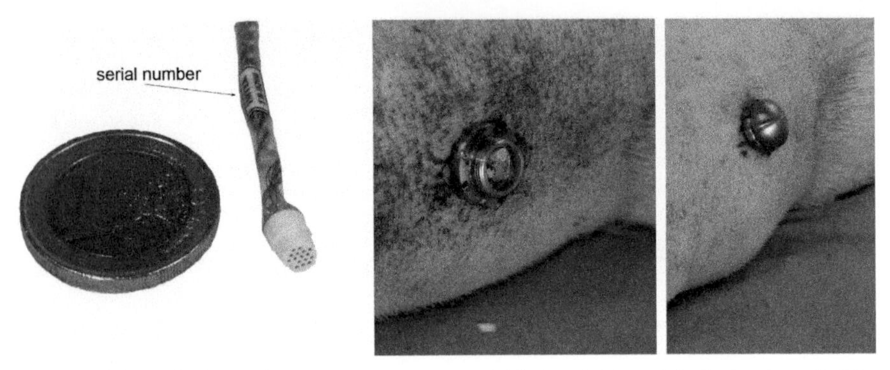

Figure 3.24 Left: Omnetics nano connector assembled to 12 strand helical cable. Middle: Metal housing holding two Omnetics connectors sutured to the skin. Right: Closed metal housing after experiments.

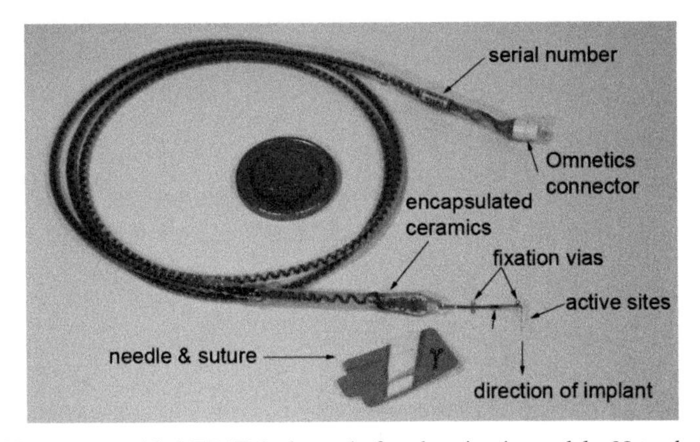

Figure 3.25 An assembled TIME-3 electrode for chronic pig models. Note that the cable length is shortened within this picture.

up to three TIME-3 electrodes in parallel (Figure 3.24 middle). When no experiment takes place, the housing can be sealed with a metal cap to avoid dirt and fluid intrusion (Figure 3.24 right). This cap will not be necessary in human clinical trials since humans can be advised not to pull on the cables that pass their skin. Human implantations with percutaneous cables have been successfully performed in the last years in Europe (Rossini 2010) as well as in the USA (Dhillon 2005).

3.4.4 Final Assembly of the TIME-3 Implants

The final TIME-3 design for the large animal model of the pig has been adapted to the pig nerve and implantation site geometry with respect to electrode pitch and cable length (Figure 3.25, Table 3.7).

Electrodes were assembled with cables and connectors, cleaned as described above, packaged in sterile bags, and sent to the partners to get sterilized before implantation in their respective environment. Serial numbers close to the connectors allow identification of electrodes even after being implanted.

Table 3.7 Specifications of assembled TIME-3 electrodes for chronic implants in pigs

	Value	Comment
Electrode		
Type	TIME-3	Design details in Figure 3.20
Ø active site	80 μm	
Number of active sites	6/12	Per side/overall
Pitch of active sites	440 μm	Covering 2.4 mm depth from the nerve surface
Indifferent electrode	Yes	One per side
Reference electrode	Yes	GND; one per side
Coating	SIROF	All active sites
Cable		
Length	65 cm	
Diameter	2.6 mm	Outer diameter
Strands	12	Helical
Strand material	MP35N	With PEI insulation
Tubing	Silicone	Without contrast strip
Connector		
Type	Omnetics NCP	Nanoplastic circular connector
Number of contacts	16	Only 12 are used for pigs
Ø connector	4.6 mm	
Length of Omnetics connector	5.6 mm	
Cleaning and Packaging		
Rinsing	70% ethanol	Complete device in cleanroom
Drying	~10 min	In cleanroom
Disinfection	Miele disinfector G7735	Certified program
Packaging	Sterile pouches	Double bagged

3.5 Life-time Estimation of TIMEs for Human Clinical Trials

Bringing new developments from preclinical trials on animal models into a first-in-human clinical trial requires verification that the components as well as the whole implant stay intact and functional over the predicted life time. In a first-in-human clinical trial, 30 days implantation time are scheduled from European laws as longest time to prove safety of the device. Therefore, evaluations have to take place to investigate and assess the stability of the different materials and components as well as the whole implant. In the case of the TIME-3H (H stands for human application), we had to assess the stability of the polyimide substrate, the stability of the electrode material with respect to stimulation pulses, and the wire-based cable with respect to stretching and bending. All tests were performed in vitro in physiologic saline solution to mimic the implantation environment. Details on the final TIME-3H design and process technology can be found in Section 3.6.

3.5.1 Lifetime Estimation of Polyimide

Although older literature (Campbell 1998, Murray 2004) reported data regarding the degradation of polyimide under high temperature and exposed to humidity, little is known about the long-term behavior of polyimide in body tissue. Our group performed long-term investigation of the mechanical properties of polyimide to evaluate its lifetime (Rubehn 2010). After storing the material in phosphate-buffered saline (PBS) at 37°C, the mechanical properties were examined with tensile test and weighing test. For accelerated ageing, the specimens were also stored at elevated temperature (60°C and 85°C) and also tested, even though 80°C are too high for polyimide to follow the Arrhenius approach for lifetime testing.

For the tensile test, polyimide specimens were designed according to ASTM D882-02 (ASTM 2002) but the dimensions were scaled down. The same kind of polyimide (U-Varnish-S, UBE) and curing process which are used in TIME-3H electrode fabrication was chosen. The Young's modulus of the polyimide film showed stable behavior over 18 months stored at 37°C (Figure 3.25, green line), while the moduli changed in samples stored at elevated temperatures (Figure 3.26, blue and red line). Other mechanical properties such as stress at break, strain at break, stress at 10% strain, and fracture energy are not changed significantly during the same time period.

To observe material degradation, gravimetric analysis was carried out. A standard polyimide foil (Upilex25S, UBE) was purchased from the manufacturer. Upilex25S is the trade name of a foil made of the precursor resin

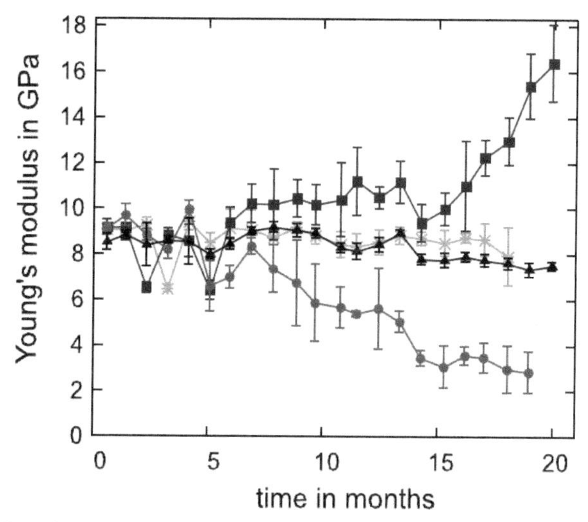

Figure 3.26 Young's modulus' change of polyimides (U-Varnish-S, UBE) stored in PBS at different temperatures.

Source: Reprinted from Rubehn, B., and Stieglitz, T. In vitro evaluating of the long-term stability of polyimide as a material for neural implants, *Biomaterials*, 31(13): 3449–3458, 2010, with permission from Elsevier.

U-Varnish-S. The mass of polyimide film stored in PBS at 37°C (Figure 3.27, triangles) did not change during 16 months.

Based on these long-term results of polyimide stored in PBS, we predicted that the lifetime of polyimide is much longer than 30 days which is the planned human trial period.

3.5.2 Stability of Iridium Oxide as Stimulation Electrode Material

Sputtered iridium oxide (IrOx) was chosen as electrode coating since its reversible charge injection capacity is much higher than the one of platinum (Cogan 2008). The lifetime of IrOx, the conductive coating on the TIME-3H electrode, was estimated by pulse tests. Pulse testing was carried out with a custom made 12-channel pulse tester (Schuettler 2008), composed of improved Howland current pumps and circuits to subtract the voltage drop over the access resistance. Applied pulses were rectangular, biphasic, and charge balanced in nature with a pulse width of 200 $\mu s/phase$ and a repetition frequency of 200 Hz. The test was carried out in PBS at room temperature.

A large area stainless steel electrode served as counter electrode. Injected charge was chosen to and held constant at 20 nC/phase, and the voltage across

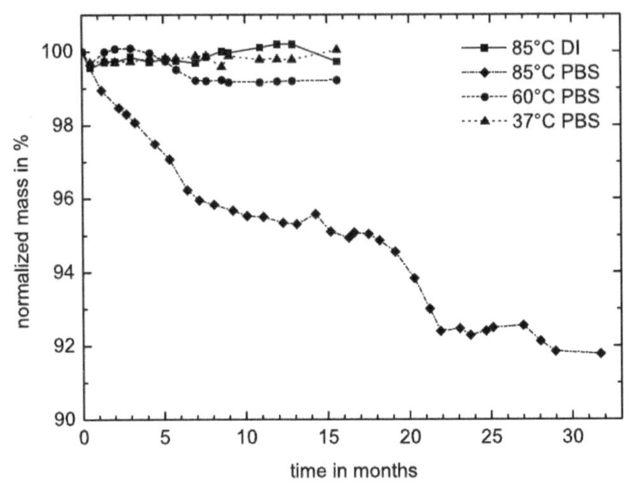

Figure 3.27 The development of the mass of polyimide films (Upilex25S, UBE) stored in PBS at different temperatures.

Source: Reprinted from Rubehn, B., and Stieglitz, T. In vitro evaluating of the long-term stability of polyimide as a material for neural implants, *Biomaterials*, 31(13): 3449–3458, 2010, with permission from Elsevier.

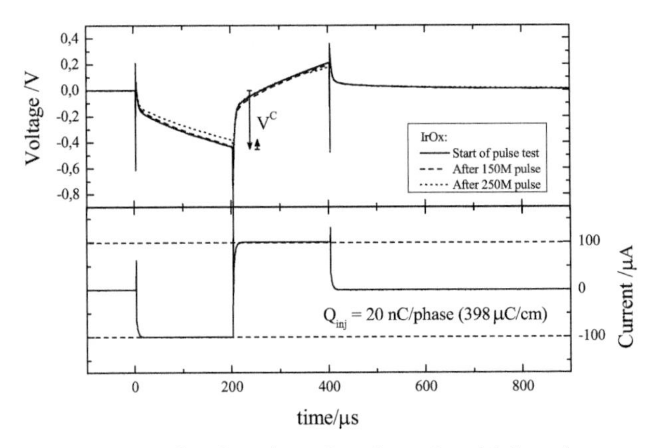

Figure 3.28 Voltage across the phase boundary (upper) and injected current pulse (lower) during the pulse test of IrOx.

the phase boundary was monitored over the timeframe of 250 million pulses (Figure 3.28). A slight decrease of the voltage across phase boundary (V_C) was observed during the 250 M injected pulses. V_C was measured as 646 ± 21 mV, 619 ± 17 mV, and 576 ± 24 mV for pristine IrOx, after 120 M pulses and after 250 M pulses, respectively. The safety limits of charge injection, which are determined by the material's water window, were not exceeded during all pulse tests.

In addition, impedance spectra were measured to quantify possible alterations during the pulse test in the IrOx coating (Figure 3.29). After the 250 M pulses, no significant change in the impedance spectra was observed, confirming that the coating is still stable.

If we consider that the absolute maximum number of pulse which will be used in the human trial is 162 M pulses (500 Hz repetition rate, 2 times of 90 min duration per day, and 30 days), these results after the 250 M pulses confirm that this IrOx coating is safe for the scheduled human trial.

3.5.3 Mechanical Stability of Helically Wound Cables

Helically wound cables connect the polyimide-based thin-film electrode via an adapter with the extracorporal stimulation equipment in the human clinical trial in a percutaneous way. These cables were made out of a helix of 16 parallel conductors (MP35N; Ø 90 μm) inside a silicone tubing. Since silicone rubbers are merely ion barriers, water would penetrate this encapsulation and might induce failure mechanisms. Therefore, the hollow tube carrying the helix is filled up with another silicone adhesive in a way that no air is entrapped within the tubing. To maintain flexibility, the shore hardness (type A) of the filler matches that of the tubing. In order to characterize

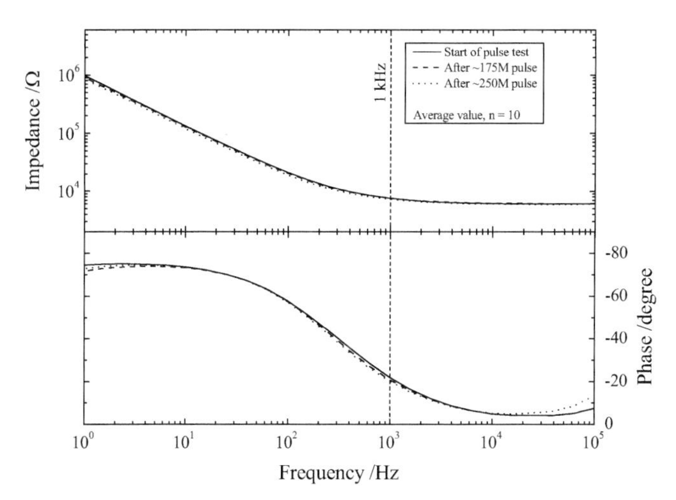

Figure 3.29 Impedance spectra of IrOx coating before pulse test, after 175 M pulses and 250 M pulses.

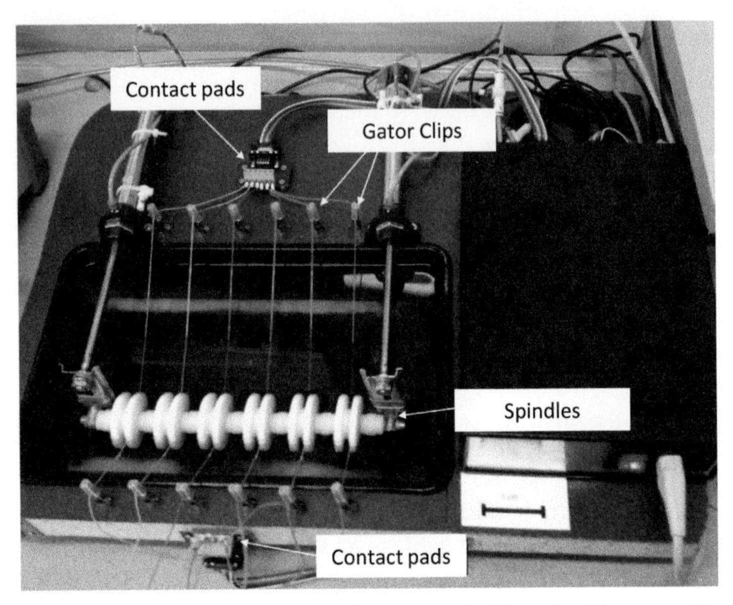

Figure 3.30 Picture of test assembly to characterize fabricated cables according to ISO 45502.

the lifetime of such a cable, a custom made testing machine was built (Figure 3.30) in accordance with the pacemaker standards (ISO 45502).

As described by the standard, the cable has to be bent by a specific bending radius, has to be inserted in the setup with a distinct load force, and has to withstand at least 47,000 cycles. Pneumatic cylinders moved a trolley of six parallel spindles, which define the bending radius (1 cm). The cables were inserted into the setup with a load force of 1 N, measured by a force meter. A sinusoidal signal with a frequency of 100 Hz and an amplitude of 1 V was applied to one conductor within each cable and used to measure impedance over the complete timeframe of the test. The trolley was moved inside a bath of saline solution (0.9% NaCl) to simulate the body environment and, in analogy, the water uptake of silicone rubber. Tests were carried out over 75,000 cycles, whereas no single cable or wire broke. Impedance values of the wires were constant over the complete timeframe of the test and varied only about ±2% from their initial value.

To further characterize the electrical properties of the cables, impedance values and coupling capacities per centimeter cable length were measured, and the coupling capacities were modeled between two adjacent conductors (Figure 3.31). While the coupling capacities to ground (C_{1G} and C_{2G}) can

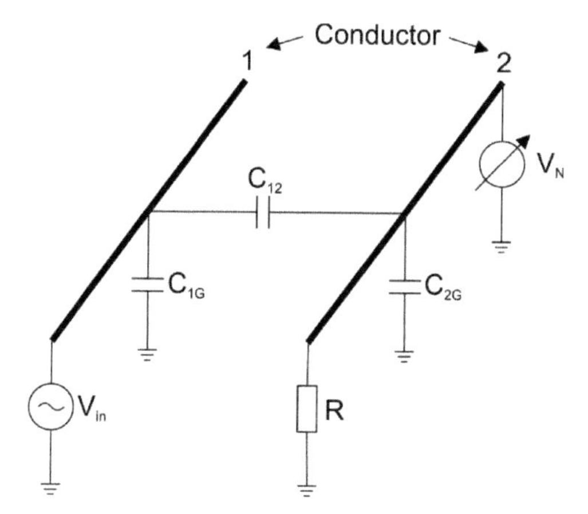

Figure 3.31 Model of coupling capacities between two adjacent conductors.

Table 3.8 Properties of helically wound cables

Property	TIME-3H Cable
Conductor material	MP35N
Material of isolation	Polyesterimide, silicone rubber hose
Outer diameter of the cable	1.80 mm
Max. usable length	105 cm
Number of conductors	16
Outer diameter of the conductor	90 μm
Resistance of the conductor	3.02 Ω cm^{-1}
Capacitive coupling to neighboring conductor	~2.6 pF cm^{-1}

be neglected within implantable cables, the "direct" coupling capacity (C_{12}) of one conductor to the other is of importance during stimulation as well as during recording. Whenever a signal is applied to conductor 1, a current will inevitably start to flow, generating an electrical field around conductor 1. This electrical field will induce a current in conductor 2 and, dependent on cable impedance and load resistor R, an electrical noise potential can be measured (V_N) in wire 2. In stimulation mode, this noise potential could, in a worst-case, induce a stimulation pulse within an active electrode site next to the targeted active site and, hence, decrease selectivity. In recording, the general noise level is increased and, again as worst-case scenario, targeted neural activity could be lost. Hence, a small coupling capacity is preferable between the two conductors.

The TIME-3H cable (Table 3.8) was tested over a period of time (~100 h) while being incubated in saline solution (0.9% NaCl). Since testing in accordance with pacemaker standards was successful, it was concluded that the

TIME cables are appropriate to be implanted for the timeframe of at least 30 days.

3.6 Requirements and Steps to Transfer Preclinical Results in Devices for the First-in-man Clinical Trial

3.6.1 Assessment of Previous Work and Pre-existing Knowledge

Assessment and documentation of previous work and pre-existing knowledge is a mandatory task in writing a so-called investigator's brochure to prove conformity with the essential requirements. Not only own work but also published studies have to be taken into account.

One clinical trial was already published in 2012 when this investigator's brochure was prepared and written. Colleagues had implanted four longitudinal intrafascicular multichannel thin-film electrodes (tfLIFE) with platinum stimulation sites in one amputee subject in December 2008 with the aim of controlling a cybernetic hand prosthesis (Rossini 2010). The fLIFE have been longitudinally implanted in median and ulnar nerves (two for each nerve) using percutaneous cables with connectors outside the body. Afferent stimulation was performed with trains of cathodal rectangular pulses with repetition rates ranging from 10 to 500 Hz, current amplitudes between 10 and 100 μA, and pulse widths between 10 and 300 μs to stay in the electrodes' safety limits. The results achieved show that 3 out of 4 implanted intrafascicular multicontact electrodes were able to elicit sensations that were related to parts of the missing hand in the fascicular projection territories of the corresponding nerves and that the sensations were modulated by varying the pulse frequency and width (Benvenuto 2010). However, the maximum allowed safe charge increased over time and sensory threshold exceeded the safe limits after 2 weeks, that is, no sensations could be further elicited. In detail, the charge threshold increased during the first 13 days from 0.1 to 2.5 nC, and no sensation had been elicited through any of the electrodes despite the 75% of maximum charge ($\sim$4 nC) at later points in time. In order to avoid irreversible electrochemical processes of the Pt electrodes, stimulation procedures were stopped. The charge increase was mainly due to fibrotic reactions around the electrode and to a progressive "habituation" of the patient moving from an initial "hypersensitivity," due to long-lasting sensory deprivation (subjective sensory threshold below maximum stimulation), then decreasing and stabilizing toward more physiological levels (subjective sensory threshold above maximum stimulation). These results made clear that a platinum coating is not sufficient to stimulate afferent fibers in humans for a longer period, due to its low charge injection capacity ($4\ \mathrm{nC} \cong 75\ \mu\mathrm{C/cm^2}$). Therefore, sputtered iridium oxide films (SIROF) were suggested and introduced as active coating, because of its comparably high charge injection capacity of about $100\ \mathrm{nC}$ ($\cong 2.3\ \mathrm{mC/cm^2}$) inside electrochemical safe regions.

Results of finite element method (FEM) modeling and ex-vivo insertion force measurements of TIME-2 devices into porcine peripheral nerves (see Chapter 4) guided the design of the human TIME-3H devices. Optimum platinum thickness for tracks and electrode sites was suggested to be at least 500 nm to let von Mises stress not exceed certain thresholds. However, this thickness could not be chosen due to process technology restrictions but platinum with iridium oxide (SIROF) was thicker than $1\ \mu\mathrm{m}$ in total. The ex-vivo measurements of TIME-2 insertions into porcine median nerve gave the maximal insertion forces and allowed to quantify the effects of geometrical parameters of the electrodes on maximal insertion forces. The aim was to limit the insertion forces necessary to transversally implant the electrode, in order to reduce the probability of breaks at the loop level (polyimide) and at the active sites level (platinum) as well as to reduce local stress and deformations of neural tissue during insertions.

Two geometrical parameters were evaluated: final electrode width ($\mathrm{W_E}$) and angle α between $\mathrm{W_{loop}}$ and $\mathrm{W_E}$ (Figure 3.32).

We transferred the results of the simulations (Chapter 4) into the electrode design. TIME-3H should have sufficient material strength for implantation into the nerve by designing the width of the substrate (W_E) to $350\ \mu\mathrm{m}$, the loop length (L) to 8 mm, and the loop width ($\mathrm{W_{loop}}$) to $100\ \mu\mathrm{m}$.

Final dimensions of the geometrical shape of TIME-3H were derived from human cadaver studies at the Universitat Autónoma de Barcelona. Five human median nerves were carefully dissected following the fascicles of the nerve trunk innervating the fingers/hand and the fascicles innervating the tributary muscles at the forearm. The number of fascicles dissected at about 5 cm proximal to the elbow was counted. Cross-section samples of the

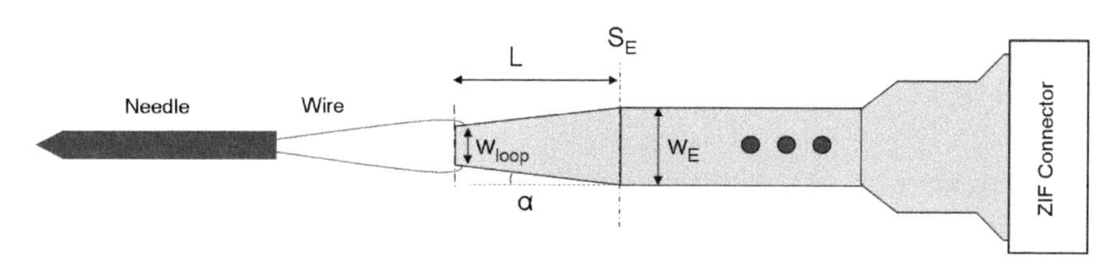

Figure 3.32 Schematic view of polyimide loop parameters to be optimized.

nerves were histologically processed at 5 and 10 cm proximal to the elbow for light-microscopy examination. Nerve diameters were in the range between 2.4 and 4.6 mm and contained between 8 and 17 fascicles. Proximal to the elbow, the fascicles innervating the hand and the muscular fascicles are not intermingled. The fascicles innervating the muscles of the forearm tend to remain at the dorsal aspect of the median nerve surrounding the fascicles of the hand, which are mostly formed by sensory fibers. Diameters were similar or larger than those of pig nerves (2.9 ± 0.4 mm) from large animal studies. Numbers of fascicles were less in human than in pig nerves (36 ± 4). Insertion forces were evaluated in the pig experiments with different custom-made (tungsten, ~75 to 100 μm diameter, ~20° tip angle) and commercially available stainless steel needles (Ethicon, 125 μm diameter, ~26° tip angle) with a pre-swaged filament. The higher diameter of the Ethicon needle causes a slightly higher insertion force needed to place it inside the nerve; however, from a mechanical point of view, there appear to be no difference between the two. The Ethicon needle was preferred for the human experiments since it is commercially available for human use.

Acute and chronic implantations in the pig forearm nerves improved the implant in the design phase. Three flaps with anchoring holes have been included in the latest design (TIME-3) to secure adequate device fixation opportunities within the nerve. The change from 14 single leads of Cooner wires (stainless steel) to helically wound MP35N cables in a single silicone rubber hose made handling during surgical intervention much easier due to the flexibility of the cables.

3.6.2 Final Electrode Design and Fabrication Technology for Human Use

The electrode design for the human clinical trial (TIME-3H) was derived from the pig version (TIME-3). Since TIME-3 electrodes were easy to handle, to fixate to surrounding tissue, and to implant transversally through the nerve, the overall design was transferred to TIME-3H electrodes and only adjusted to the larger diameter of human median nerves (Figure 3.33, Table 3.9) and the number of fascicles. Process parameters were kept unchanged for the complete fabrication and assembling technologies, since these were successfully tested in vitro and in vivo. Due to this larger nerve diameter (~4 mm in humans compared to ~2.5 mm in pigs), two additional active electrode sites could be implemented by increasing the electrodes width by merely 70 μm (from 280 to 350 μm). In comparison, the nerve diameter is enlarged by a

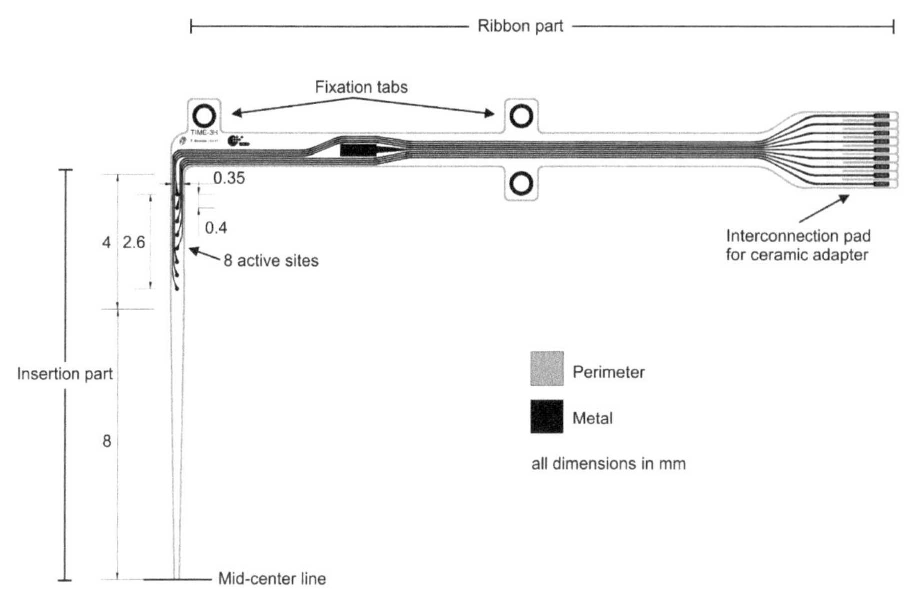

Figure 3.33 Schematic design of TIME-3H.

Table 3.9 Specifications of TIME-3 and TIME-3H designs

	TIME-3 (Pig Model)	TIME-3H
	Insertion part	
Number of active sites	6 per side/12 total	8 per side/16 total
Pitch of active sites	440 μm	400 μm
Ø of active sites	80 μm	80 μm
Coating	SIROF	SIROF
Charge injection capacity	$2.3\ \mathrm{mC/cm^2}$	$2.3\ \mathrm{mC/cm^2}$
Impedance	6 kΩ @ 1 kHz	6 kΩ @ 1 kHz
Length of insertion part	2.2 mm	4 mm
Width of insertion part	280 μm	350 μm
Loop length	5 mm	8 mm
Loop width at folding point	100 μ m	100 μm
	Ribbon part	
Width (without tabs)	1 mm	1 mm
Length	22 mm	22 mm
Number of tabs	3	3
Diameter of tabs	500 μm	500 μm
Dimension of ground electrode	1 mm × 250 μm	1 mm × 250 μm

Note: Only dimensions are changed to fit the larger human nerve. Process parameters remained unchanged.

factor of 1.6, while the electrodes width is increased by only a factor of 1.25. To keep the insertion forces still low, the loop length was increased by 3 mm from 5 to 8 mm, which is in accordance with recommendations from simulations (see Chapter 4). Since this part of the loop will not remain inside the nerve but merely acts as a guiding structure of the device during implantation, it will not influence the structural biocompatibility of the device. An increased thickness of the platinum layer, also a recommendation from simulations, could not be implemented because of technological restrictions during metal deposition. The deposition of iridium oxide (SIROF), however, results in a total layer thickness of $>$1,000 nm (300 nm Pt + 800 nm SIROF) and should therefore meet the requirements. These electrodes with this Pt-SIROF sandwich were those already evaluated in chronic rat and pig experiments and proved to be stable and biocompatible.

Manufacturing of the TIME-3H remained unchanged (see Section 3.2.1) to obtain the devices with platinum tracks sandwiched between two polyimide layers and electrodes coated with SIROF. The produced devices were left on the wafers to get introduced into the quality management (QM) of the laboratory. Batch numbers were assigned and series of tests were applied to ascertain reproducibility and quality requirements.

Cable, connector, and ceramic adapters were assembled. This complete part is consigned to the following protocol to ensure an overall immaculate surface (Donaldson 2008), since all the steps performed later, given as follows, are conducted in a class 10000 cleanroom:

1. Removal of flux residues from soldering with flux-remover (de-flux 160 by Servisol, Iffezheim, Germany)
2. Removal of flux-remover residues by washing with isopropyl alcohol (IPA)
3. Prevent IPA residues by washing in deionized water
4. Slew device five minutes in Leslie's Soup[1]
5. Removal of residues by rinsing with deionized water
6. Wash in deionized water till the conductance value of DI water settles again
7. Dry on lint-free cloth at 60°C in an oven for at least 3 hours.

The thin-film electrode was then microflexed to the ceramic adapters (Meyer 2001). In the next step, the needle with attached loop-thread (STC-6, Ethicon, Johnson & Johnson GmbH, Norderstedt, Germany) was inserted

[1]Leslie's Soup is named after a British researcher and consists of the glass cleaner "Teepol-L" (0.5%), Na_3PO_4 (2.5%), and deionised water (97%).

into the insertion part of the electrode, and the ceramic adapters were placed back to back together and glued via a droplet of silicone rubber. Following the curing period, which hardens the polymer, the ceramic adapters were encapsulated in medical grade silicone (MED-1000, PolyTec GmbH, Waldbronn, Germany). Special attention is hereby dedicated to ensure a void-free film, firstly, by applying, a swelled silicone tubing around the ceramic adapter sandwich and, secondly, by filling up the complete tube with silicone rubber. Within these processes, each batch and wafer receives a unique batch number and can, henceforth, be tracked individually. Single device of the wafer was assembled with the ceramic adapters, and the helically wound cables and individual series numbers were assigned. The ceramic adapters, Omnetics connectors, and electrode devices can be traced back by their corresponding batch number; devices were identified by their serial number on the cable.

After the device is completely assembled, it is again cleaned and sterilized according to the following protocol:

1. Transfer device into dedicated washing bag
2. Slew in ethanol for 5 minutes
3. Dry on lint free cloth for 10 minutes without washing bag
4. Transfer device into implant washing machine (Miele Disinfector G7735) for cleaning and disinfection
5. Transfer device into a sterile pouch (double packed) and label accordingly
6. Steam sterilize packed pouch at 2 bar and 134°C for 20 minutes (prion program) using an autoclave (Superior 24B, Medesign GmbH, Holzwickede, Germany).

The cleaned, packaged, and sterilized devices were ready for implantation in a human clinical trial.

3.6.3 Quality Management System

The main idea behind a quality management system (QMS) is to be able to relate any given product or product group to the single part it is made of and, hence, how this specific part was manufactured and to trace back possible failures in vivo to the manufacturing process. In addition, risk management that accompanies the development process will minimize possible hazards to the patient. To achieve this, processes, procedures, and documents have to be created, implemented, and controlled. They are hierarchically ordered and comprise all process steps and parts needed for manufacturing.

In 2010, the Laboratory for Biomedical Microtechnology at the University of Freiburg (Germany) received accreditation for its laboratory for electrode manufacturing (LEF) to manufacture medical devices according to ISO 13485:2007 (certificate number CQ103013-13). A general documentation scheme was established, reviewed, and finally certified. The quality management is currently becoming updated to the actual ISO 13485:2016 requirements.

Development of devices in accordance with such a QMS needs to cover many aspects during different phases, from the idea to a clinical trial.

3.6.3.1 Documentation of device development

Device development has to be done in close collaboration with the clinical partners from the very first idea. Discussion has to be documented to fix decisions and document target specifications. Our QMS followed the four-stage development model. Since we did not intend to apply for CE mark but perform an investigator-initiated clinical trial to address research hypotheses, we did not completely develop a device dossier but preferred a lean technical documentation to comply with the essential requirements and meet the needs of the responsible ethical committee and the legal authorities in Italy. Therefore, an investigator brochure was written containing all the available information of the design and development process including results from literature and preclinical studies. A user manual for the TIME system was written including specifications like system dimensions, electrode impedance, maximal possible charge injection capacity, and a guide for handling the TIME system. This manual was delivered with each electrode.

3.6.3.2 Risk management

Possible risks and measures to minimize potential risks in critical points have been identified and were handled according to the procedures described in the ISO 14971:2009. We used risk analysis and FMEA to assess and manage risks in the design and manufacturing part of the implantable device. Medical partners contributed to the risk assessment during handling, surgery, and implantation/explantation period.

3.6.3.3 Quality management system for device manufacturing

Manufacturing instructions for the TIME-3H have been written and implemented (Figure 3.34). For each assembly group or product, a working order with a unique number has to be written. This working order contains three additional documents: (1) the working plan, which specifies the single manufacturing steps to be carried out and records the performing person; (2) the progress procedure, which gives a detailed description of the single manufacturing steps according to the working plan; and (3) the list of parts, which lists all parts used within the manufacturing and records all batch and/or LOT numbers. Through this documentation scheme, it is always possible to trace back: the materials used, the procedure used to manufacture this specific part, and the person who manufactured it. Furthermore, fabrication on stock becomes possible, for example, for assembly groups, since all working numbers are unique and retraceable.

The TIME implant can be divided into five general assembly groups (AG), whose fabrication is monitored via the above-mentioned procedures (Figure 3.35).

On the lowest level, the AGs "ceramic adapter" and "helix cable and tubing" are situated. Here, the ceramic adapters are manufactured via screen printing of thick films onto ceramic substrates and labeled according to the batch number. Similarly, the helix cable and tube are assembled as stock articles and labeled accordingly. With these two assembly groups, the cable itself can be built up and cleaned (AG Cable). Since the electrode itself, that is, the polyimide thin film, is manufactured within the micromachining cleanroom (without quality management system), it has to be introduced into the QMS (AG Electrode). Therefore, an entry control was established that

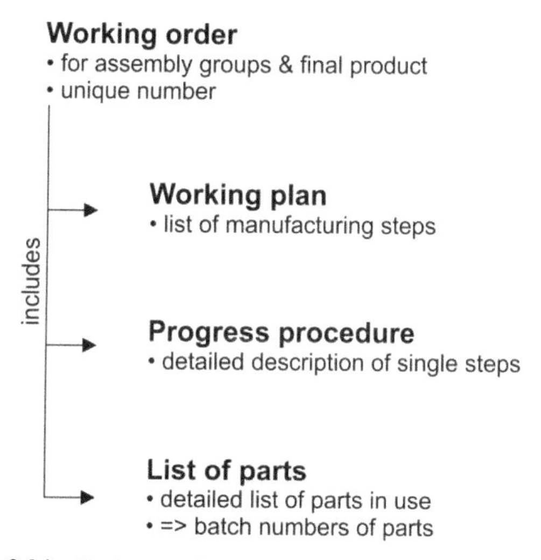

Figure 3.34 Basic tree of documentation according to ISO13485.

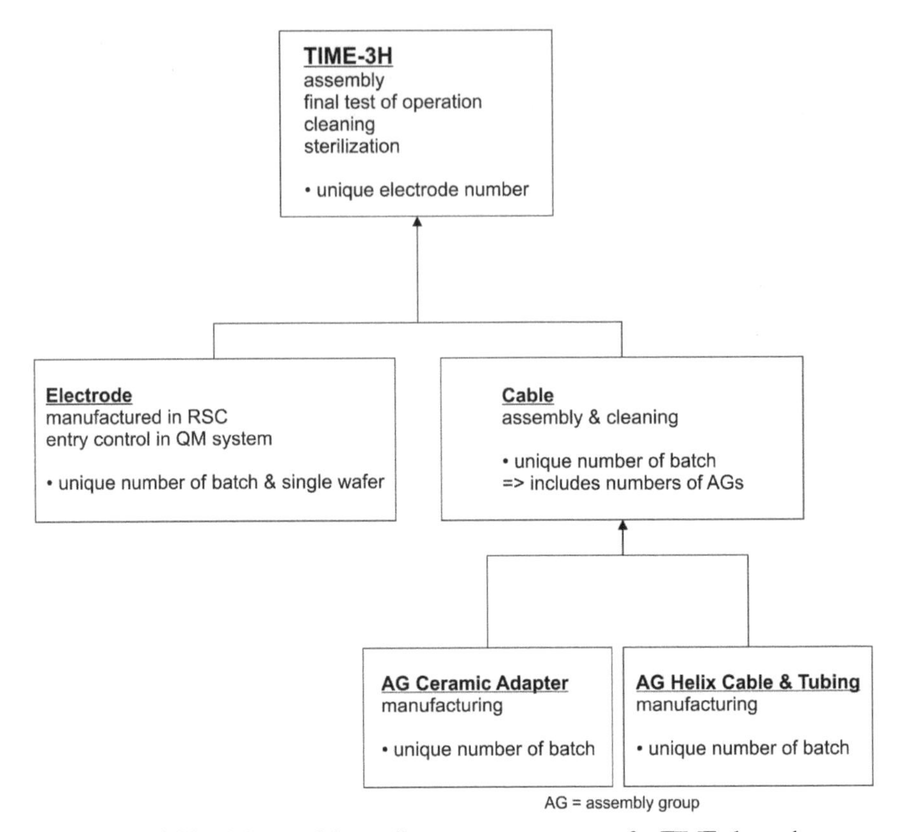

Figure 3.35 Scheme of the quality management system for TIME electrodes.

includes specific tests to reassure and maintain the quality of the polyimide structures. Included tests are Fourier transform infrared spectroscopy (FTIR) to control the amount of solvent within the polyimide layer, electrochemical impedance spectroscopy (EIS) to assert that the metal coating is stable, and optical characterizations to ensure pattern accuracy.

As a last step, the complete TIME system can now be assembled (AG TIME-3H). Here, the "AG Electrode" and "AG Cable" are joined together and encapsulated with silicone rubber. Thereafter, the electrode system is tested for functionality and subsequently cleaned and disinfected.

This final test of operation is done via electrochemical impedance spectroscopy (EIS), whereas specific limits for electrode impedance have to be met, and pulse testing, in which the limit of maximal injectable charge has to meet minimally the defined value within the electrodes manual. Disinfection of the electrode is achieved by washing the TIME system in a certified machine (Desinfektor G7735 by Miele AG, Gütersloh, Germany) using different disinfection agents. Afterwards, the complete TIME system becomes double-bagged in sterile pouches, and labeled and sterilized in an autoclave (Superior 24B by Stärker-MEDESIGN GmbH, Holzwickede, Germany) using the prion program (135°C, 3 bar, 20 min).

The resulting TIME system has a unique serial number with which all the above-mentioned steps can be retraced and material batches identified. Labeling of sterile devices will be done according to EN 554, 556 (sterilization with moist heat) and EN 868 (packaging materials and systems for medical devices). The system will be shipped (or brought in person) with the accompanying documents like electrode manual and results of the final test of operation to the hospital in which the implantation takes place.

3.7 Discussion

It has been a long way from first ideas of a transversal intrafascicular multichannel electrode (TIME) array to an implantable device that has the potential to investigate new treatment options in human clinical trials. A lot of experience has been gained in close collaborations with partners in a multidisciplinary consortium on the impact of design changes on stability and selectivity, biocompatibility and foreign body reactions, selectivity and stability of the whole system and matching electrode and stimulator performance. The reward has been the transfer of a batch of samples to clinical partners for a first-in-man clinical trial and its success. Many lessons have been learned to get an implant stable over months and more will have to be learned to get it stable over years and decades. However, at a certain point in the development phase, one has to take the responsibility to enter first-in-human clinical trials since only humans can give feedback whether the stimulation is useful and pleasant to be integrated in the activities of daily life. For one subject in one trial with four TIME being implanted and another four being the "backup safety solution", all in all more than 100 devices had to be manufactured to validate manufacturing and cleaning procedures as well as sterilization. This translational research needed a long breath and showed the huge potential of microsystem technologies and the TIME concept. The longest part of the road, however, is still ahead as such systems are available as medical devices in clinical practice and get reimbursed by general health insurances.

References

Benvenuto, A., Raspopovic, S., Hoffmann, K. P., Carpaneto, J., Cavallo, G., Di Pino, G., Guglielmelli, E., Rossini, L., Rossini, P. M., Tombini, M. and Micera, S. (2010). Intrafascicular thin-film multichannel electrodes for sensory feed-back: evidences on a human amputee, 32nd Annual International Conference of the IEEE Engineering in Medicine and Biology Society, Buenos Aires, Argentina, 1800–1803.

Boretius, T., Zimmermann, D. and Stieglitz, T. (2009). "Development of a Corrugated Polyimide-Based Electrode for Intrafascicular Use in Peripheral Nerves", IFMBE Proceedings 25/IX, Springer, Munich, Germany, pp. 32–35.

Boretius, T., Badia, J., Pascual-Font, A., Schuettler, M., Navarro, X., Yoshida, K. and Stieglitz, T. (2010). "A transversal intrafascicular multichannel electrode (TIME) to interface with the peripheral nerve", Biosens Bioelectron, 26(1): 62–69.

Bossi, S., Menciassi, A., Koch, K. P., Hoffmann, K. P., Yoshida, K., Dario, P. and Micera, S. (2007). "Shape Memory Alloy Microactuation of tf-LIFEs: Preliminary Results," IEEE Trans. on Bio-Med Eng, vol. 54, no. 6, pp. 1115–1120.

Bossi S., Kammer, S., Doerge, T., Menciassi, A., Hoffmann, K. P., Micera, S. (2009). "An Implantable Microactuated Intrafascicular Electrode for Peripheral Nerves," IEEE Trans. on Bio-Med Eng, vol. 56, no. 11, pp. 2701–2706.

Brummer, S. B. and Turner, M. J. (1977). "Electrical Stimulation with Pt Electrodes: I – A Method for Determination of "Real" Electrode Areas," IEEE T Bio-Med Eng, vol. BME-24, no. 5, pp. 436–439.

Campbell, F. J., Brewer, A. K., Orr, R. J., Janicke, T. A. and Bruning, A. M. (1998). "Hydrolytic Deterioration of Polyimide Insulation on Naval Aircraft Wiring," Annual Report: Conference on Electrical Insulation and Dielectric Phenomena, Oct. 16–20, Ottawa, Ont., Canada. pp. 180–188.

Cogan, S. F. (2008). "Neural Stimulation and Recording Electrodes," Annu Rev Biomed Eng, vol. 10, pp. 275–309.

Cogan, S. F., Troyk, P. R., Ehrlich, J., Gasbarro, C. M. and Plante, T. D. (2007). "The Influence of Electrolyte Composition on the In Vitro Charge-Injection Limits of Activated Iridium Oxide (AIROF) Stimulation Electrodes," J Neural Eng, vol. 4, no. 2, pp. 79–86.

Craggs, M. D., Donaldson, N. d. N. and Donaldson, P. E. K. (1986). "Performance of Platinum Stimulating Electrodes, Mapped on the Limit-Voltage Plane. Part 1: Charge Injection In Vivo," Med Biol Eng Comput, vol. 24, pp. 424–430.

Dhillon, G. S. and Horch, K. W. (2005). "Direct Neural Sensory Feedback and Control of a Prosthetic Arm," IEEE T Neural Network Rehabil Eng, vol. 13, no. 4, pp. 468–472.

Donaldson, N. d. N. and Donaldson, P. E. K. (1986). "Performance of Platinum Stimulating Electrodes, Mapped on the Limit-Voltage Plane. Part 2: Corrosion In vitro," Med Biol Eng Comput, vol. 24, pp. 431–438.

Donaldson, N. de N. (2008). Personal communication.

Franks, W., Schenker, I., Schmutz, P. and Hierlemann, A. (2005). "Impedance Characterizationand Modelling of Electrodes for Biomedical Applications," IEEE T Bio-Med Eng, vol. 52, no. 7, pp. 1295–1302.

Kovacs, G. T. A. (1998). Micromachined Transducers Sourcebook, WCB McGraw-Hill, Chapter 3.6 Mechanical Actuators.

Kue, R., Sohrabi, A., Nagle, D., Frondoza, C. and Hungerford, D. (1999). "Enhanced proliferation and osteocalcin production by human osteoblast-like MG63 cells on silicon nitride ceramic discs," Biomaterials, vol. 20, no. 13, pp. 1195–1201.

Liu, X., Demosthenous, A. and Donaldson, N. d. N. (2008). "Platinum electrode noise in the ENG spectrum," Med Biol Eng Comput, vol. 46, no. 10, pp. 997–1003.

Marrese, C. A. (1987). "Preparation of Strongly Adherent Platinum Black Coatings," Anal Chem, vol. 59, pp. 217–218.

Meyer, J.-U., Stieglitz, T., Scholz, O., Haberer, W., Beutel, H. (2001). "High Density Interconnects and Flexible Hybrid Assemblies for Active Biomedical Implants," IEEE Trans. on Advanced Packaging, vol. 24, no. 3, pp. 366–374.

Murray, S., Hillman, C. and Pecht, M. (2004). "Environmental Aging and Deadhesion of Polyimide Dielectric Films," Journal of Electronic Packaging, vol. 126, no. 3, pp. 390–397.

Neumann, A., Reske, T., Held, M., Jahnke, K., Ragoss C. and Maier, H. R. (2004). "Comparative investigation of the biocompatibility of various silicon nitride ceramic qualities in vitro," J. Mater. Sci. – Mater. M., vol. 15, no. 10, pp. 1135–1140.

Robblee, L. S. and Rose, T. L. (1990). "Electrochemical guidelines for selection of protocols and electrode materials for neural stimulation," in

Neural Prostheses – Fundamental Studies, 1 ed. W. F. Agnew and D. B. McCreery, Eds. Englewood Cliffs, New Jersey: Prentice Hall, pp. 25–66.

Rossini P. M., Micera, S., Benvenuto, A., Carpaneto, J., Cavallo, G., Citi, L., Cipriani, C., Denaro, L., Denaro, V., Di Pino, G., Ferreri, F., Guglielmelli, E., Hoffmann, K. P., Raspopovic, S., Rigosa, J., Rossini, L., Tombini, M. and Dario, P. (2010). "Double Nerve Intraneural Interface Implant on a Human Amputee for Robotic Hand Control," Clinical Neurophysiology.

Rubehn B. and Stieglitz, T. (2010). "In Vitro Evaluating of the Long-Term Stability of Polyimide as a Material for Neural Implants," Biomaterials, vol. 31, no. 13, pp. 3449–3458.

Schuettler M., Franke, M., Krueger T. B. and Stieglitz, T. (2008). "A Voltage-Controlled Current Source with Regulated Open Circuit Potential for Safe Neural Stimulation," Journal of Neuroscience Methods, vol. 171, no. 2, pp. 248–252.

Schuettler, M. (2007). "Electrochemical Properties of Platinum Electrodes in Vitro: Comparison of Six Different Surfaces Qualities," Proc of the 29th Ann Internat Conf of the IEEE EMBS, p. 4.

Schuettler, M., Stieglitz, T. (2013). Assembly and Packaging. In: Hodgins D., Inmann, A. (Eds.): Intelligent Implantable Sensor Systems for Medical Applications. Cambridge: Woodhead Publishing Ltd., pp. 108–149.

Schuettler, M., Doerge, T., Wien, S. L., Becker, S., Staiger, A., Hanauer, M., Kammer, S. and Stieglitz, T. (2005). "Cytotoxicity of Platinum Black," Proc 10th Ann Conf of the IFESS 2005, pp. 343–345.

Shannon, R. V. (1992). "A Model of Save Levels for Electrical Stimulation," IEEE T Bio-Med Eng, vol. 39, no. 4, pp. 424–426.

Stieglitz, T. (2004). "Materials for Stimulation and Recording – Electrode Materials for Recording and Stimulation," in Neuroprosthetics – Theory and Practice. Ed. 1. K. W. Horch and G. S. Dhillon, Eds. Singapore: World Scientific Publishing Co. Pte. Ltd., pp. 475–516.

Stieglitz, T., Schuettler, M., Rubehn, B., Boretius, T., Badia, J., Navarro, X. (2011). Evaluation of Polyimide as Substrate Material for Electrodes to Interface the Peripheral Nervous System. Proc. of the IEEE-EMBS Neural Eng Conf., April 27th-May 1st, 2011, Cancun, Mexico, pp. 529–533.

Veraart, C., Duret, F., Brelen, M., Oozeer, M. and Delbeke, J. (2004). "Vision Rehabilitation in the Case of Blindness," Expert Rev Med Devices, vol. 1, no. 1, pp. 139–153.

Zhou, D. M. (2005). "Platinum Electrode and Method for Manufacturing the Same," United States Patent US 6,974,533 B2.

4

Modeling to Guide Implantable Electrode Design

Giacomo Valle[1,2] and Silvestro Micera[1,2,*]

[1]Bertarelli Foundation Chair in Translational Neuroengineering, Centre for Neuroprosthetics and Institute of Bioengineering, School of Engineering, École Polytechnique Fédérale de Lausanne (EPFL), Lausanne, Switzerland
[2]The BioRobotics Institute, Scuola Superiore Sant'Anna, Pisa, Italy
E-mail: silvestro.micera@epfl.ch
*Corresponding Author

Neuroprostheses are becoming widespread clinical solutions, addressing human nervous system at different levels. This technology can significantly improve the quality of life of people who have suffered from different neurological disabilities. Despite the large number of peripheral nervous system (PNS) electrodes available and their good performances, the ever-growing complexity of neuroprosthetic devices trying to mimic the natural hand implies a constant need to improve electrode selectivity. This is particularly true for stimulating electrodes whose aim is to mimic the natural sensory feedback from the hand arising from a very dense network of afferents serving different modalities (especially in the fingertips) by only stimulating at discrete, restricted locations on a given nerve (Riso, 1999). Hence, an important goal for a PNS electrode is to achieve the highest selectivity for a high number of nerve fascicles while minimizing the invasiveness and potential nerve damage. In this context, experimental studies have been conducted in order to compare the selectivity performances of different types of PNS electrodes (Badia et al., 2011). Animal models are common developmental tools for testing peripheral nerve interfaces. However, the complexity of the nerve tissue upon which stimulation is applied, as well as the anatomical differences between animal models and humans, induce great variability regarding the neural response (Grinberg et al., 2008). Furthermore, the wide range of design factors that can influence the outcome of the stimulation, such as electrode type and position or the stimulation pattern (amplitude, pulse width, frequency, monopolar, or multipolar stimulation), needs to be explored in order to optimize a stimulation protocol and the neural interfaces for a given application, thus requiring a large number of experimental trials and subjects. Even so, the interaction with living tissue induces an inevitable variability in experimental results due to several factors that cannot always be identified, thus rendering the problem even more complicated.

The use of computer models to study the electrical stimulation of neural systems appears to be an inexpensive and efficient way to tackle this issue and thus assist in the development of neural devices or applications, by exploring the high dimensional space of design parameters while minimizing animal use. Among the first explorations of an influence of external electrical fields on the neurons, by analytical modeling, was performed by McNeal (1976), who developed the concept of so-called "activating function." He used a fact that even though neural devices and neurons have different communicating currents, they both share same electric field. The modulation of it by the injected electrical currents can depolarize the external membrane of neurons, provoking the ionic currents flows, and finally the generation of spikes, which are the basic carriers of information in the human nervous system. Activation function is proposing that the likelihood of neural activation by external stimulation is proportional to the second derivative of external field respect to neuronal spatial extension. This idea is extended and analytically improved in Rattay's works (1986, 1989), which extended the concept from the point sources of current to the realistic, similar-to-electrode sources. Although the activation function is yet used as a most rapid and intuitive indicator of the approximate estimation of axonal responses to electrical stimuli, the recent works (Zierhofer, 2001; Moffitt et al., 2004) have shown that it is introducing the important mistakes. The main reasons for these errors were that approach based on the activation function, was neglecting of high nonlinearity present in axonal answers and the realistic anisotropy of a medium in which neurons are placed. Recent computational models do account for both the anisotropic extracellular conductivity present in the nerves, and for the dynamic response of neuronal cells and axons to the extracellular electrical stimulation. For calculating the voltages induced by means of electrical stimuli injected by electrode into the anisotropic medium, the finite element method (FEM) are exploited. The estimation of the axonal responses to the external stimuli was investigated by means of software dedicated for efficient calculus of neuronal dynamic (McIntyre, 2002) and cable equations, NEURON (Hines and Carnevale, 1997). Finally, the FEM results are interpolated into the NEURON model, obtaining together what can be called a "hybrid electro-neuronal model."

Initial concept of hybrid modeling was proposed in the studies regarding the electrical epidural stimulation (EES) of spinal cord (Coburn, 1985; Coburn and Sin, 1985). Then, similar idea has been exploited in works that model extracellular stimulation of central nervous system neurons, and in particular for the purpose of deep brain stimulation (DBS) modeling (McIntyre and Grill, 2002; Miocinovic et al., 2006). In the recent past, it has been also used in human peripheral nervous system to optimize the design of extraneural cuff electrodes (Schiefer et al., 2016) for the neural stimulation in effort to make a motor rehabilitation of spinal-cord-injured patients. Intraneural electrodes have been simulated (Raspopovic et al., 2011) and validated (Raspopovicet et al., 2012) in the rat nervous implants, using the same hybrid modeling approach. Successful translational use of models in the CNS, as the development of software CICERONE (Frankemolle et al., 2010), which is used in the DBS practice, or in model for spinal cord simulation (Moraud et al., 2016), which was the fundamental for the development of sophisticated stimulation paradigms, that enabled unseen level of mobility in fully Spinal Cord Injured (SCI) rats. These models were not only important from the translational viewpoint: by indicating the electrodes placement and paradigm of optimal stimulation, but also enabled the deep understanding of interactions, which is the fundamental of these intervention computer simulations (Rattay et al., 2000; Capogrosso et al., 2013) provided evidence that EES primarily engages large myelinated fibers associated with proprioceptive and cutaneous feedback circuits during the SCI rehabilitation.

Pivotal role of modeling in all these complex systems, increases the need for the implementation of similar realistic models also for the human PNS, to be exploited within the neuroprostheses (Raspopovic et al., 2014; Tan et al., 2014) development. Neural interfaces are an important component of these systems, which allow direct communication with the nervous system. Several neural interfaces for the PNS have been developed during the past year. They range from epineural electrodes, having low invasiveness and low selectivity, to regenerative electrodes, having higher selectivity but at the same time, also higher invasiveness (Navarro et al., 2005). A good tradeoff between the two previous solutions can be found in intraneural interfaces (as transverse intrafascicular multichannel electrode (TIME) (Boretius et al., 2010) or self-opening intraneural peripheral interface electrode (SELINE)

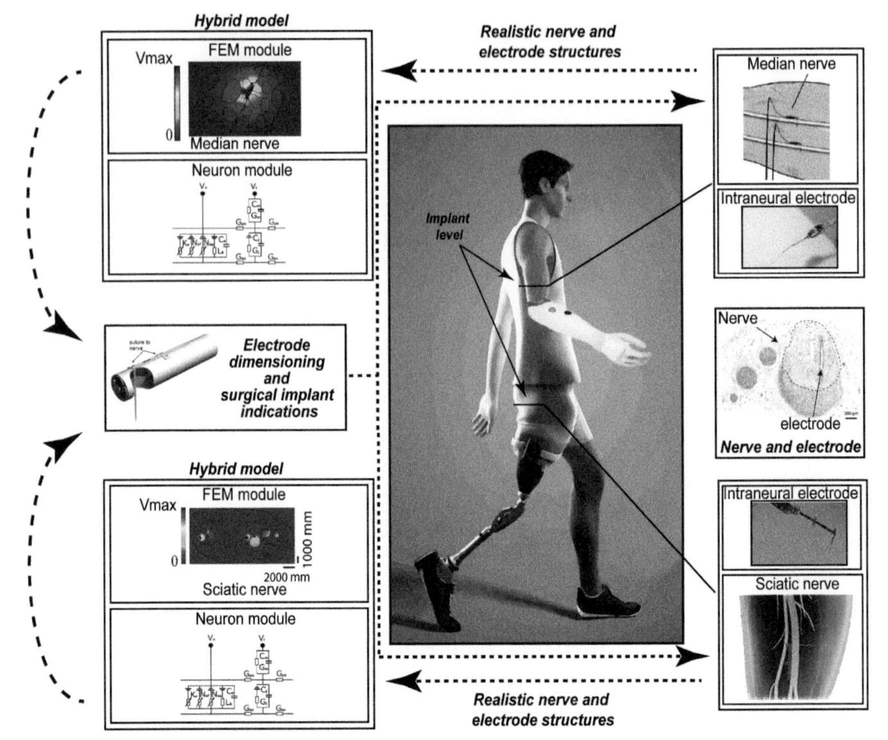

Figure 4.1 Hybrid modeling: nerves sections are taken at the appropriate level for the implantation, and then used within the hybrid electroneuronal models for the development of the optimized neural interfaces for selective, gradual, and minimally invasive use.

(Cutrone et al., 2015). Raspopovic and his collaborators (Raspopovic et al., 2017) were depicting the prominent use of models of human median and sciatic electrical nerves stimulation (ePNS) within the framework of development of innovative neuroprostheses (Figure 4.1).

The efficacy of neuroprostheses can be improved by increasing the possibility of neural interfaces used to stimulate specific subsets of neurons, while not stimulating the untargeted ones, the concept which is measured by the electrode's selectivity. Models could help to reach the scope of having devices that are enhancing selectivity, while reducing invasiveness and also decreasing the amount of current to be injected into the neural tissues. Selectivity is mainly influenced by interface design, and more in particular with dimensions of whole device, its shape, number, and distance of active contacts, used for the stimulation of the neural tissue. Models can indicate the optimal number of devices to be implanted into an individual nerve, and therefore they should give hints regarding how the neurosurgery should be performed (Raspopovic et al., 2017).

In the past years, the neural interfaces were tested in vitro or in animal preparation by single-channel stimulation and measurements of same type of output measure (Badia et al., 2011). This conceptual framework together with time limitation during the effective use, have restricted the clinical application of peripheral nerve stimulation to continuous, single active sites injected stimulation patterns. However, this strategy does not exploit optimally, all the possible capability offered by implanted devices, particularly does not allow to address subject-specific deficits which is pivotal to maximize the outcome of rehabilitation protocols. The use of more sophisticated stimulation paradigms (Fang and Mortimer, 1991; Grill and Mortimer, 1995, 1997; Vuckovic et al., 2004; Hennings et al., 2005), or combinations of single active sites stimulations into the complex multipolar stimulation, could be promising, and should be extensively explored by use of models (Schiefer et al., 2012; Saal and Bensmaia, 2015; Oddo et al., 2016; Saal et al., 2017).

Finally, among the biggest problems encountered during the use of different neural interfaces, is the temporal change of charge necessary to guarantee the therapeutic use. This is most probably due to the tissue complex reaction, and some aspects of it can be interpreted by use of models, rather than extensive animal sacrificing and following histological analysis.

Computer models can be useful for exploring the high dimensional space of design parameters with the goal to provide guidelines for the development of more efficient neural electrodes, with minimal animal use and optimization of manufacturing processes.

4.1 Hybrid Model

The use of computer models to study the electrical stimulation of neural systems appears to be an inexpensive and efficient way to assist in the development of neural devices or applications. The state of art in models accounts for anisotropy of extracellular conductivity, present in real nerves, and also for the nonlinear response of cells to the extracellular stimulation. Those two aspects are solved separately: by means of the FEM which solves the voltage distribution generated by injected currents, and by using calculations of neuronal dynamics to estimate the axonal response to the electrical stimulations. This kind of model has been called hybrid field-neuron models (or hybrid FEM/Neuron models). To couple the external electric fields with the fiber or cell, proper models to account for external stimulation were developed.

The state of the art most used approach is the so-called compartmental modeling of fibers, which is based on the subdivision of fibers and cells into elementary circuit representation used to model the different parts of the cell or fiber, like axons, somas, and nodes of Ranvier.

4.2 Finite Elements Model

As a first step, the correct heights of the nerve for implantation have to be determinate, and corresponding histological picture needs to be found. Considering upper and lower limb implants, the correct height to consider is above the elbow for the transradial (under-elbow) amputees, while the level at the ischial tuberosity for transfemoral (thigh-level) leg amputees.

Secondly, anatomically shaped geometrical model of the nerve and fascicles are segmented by using the freeware software ImageJ (by freeware software ImageJ with NeuronJ plug-in) obtaining an anatomically shaped geometrical model. Coordinates of the image segmentation are then exported to MATLAB (livelink COMSOL-MATLAB), where a 2D recreation of the nerve is constructed. Since the fascicles are surrounded with a connective tissue sheath called perineurium and it can influence the final results of potential distribution, it was crucial to separate it from the fascicles' contour. As reported in Grinberg et al. (2008) perineurium thickness being determined by the size of fascicle and it is equal to 3% of fascicle's radius. Then the coordinates were interpolated with interpolation curve. The segmented geometry is imported into the FEM software, COMSOL (COMSOL S.r.l., Italy) and extruded along the longitudinal axis achieving a 3D structure (Figure 4.2).

Very important aspect is a correct assignment of different electrical values to the separated tissue classes: epineurium, perineurium, and endoneurium. These values are available from literature, however, need to be critically revision, and adapted to the particular model. Indeed, the intrafasicular endoneurium, debt to the longitudinal disposal of axon within, has an anisotropic conductivity tensor with a longitudinal value of 0.571 S/m and a transverse value of 0.0826 S/m. The epineurium is assumed to be an isotropic medium with a conductivity of 0.0826 S/m (Schiefer et al., 2008). The perineurium is modeled as an isotropic conductor taking into account the thickness of the perineurium as 3% of the approximately diameter of the fascicle (Grinberg et al., 2008), and the difference temperature between frog and humans, with a value of 0.00088 S/m (Raspopovic et al., 2017). Generally, the surroundings of nerves are implemented as homogeneous saline solution (2 S/m), which is emulating the intraoperative environment, with

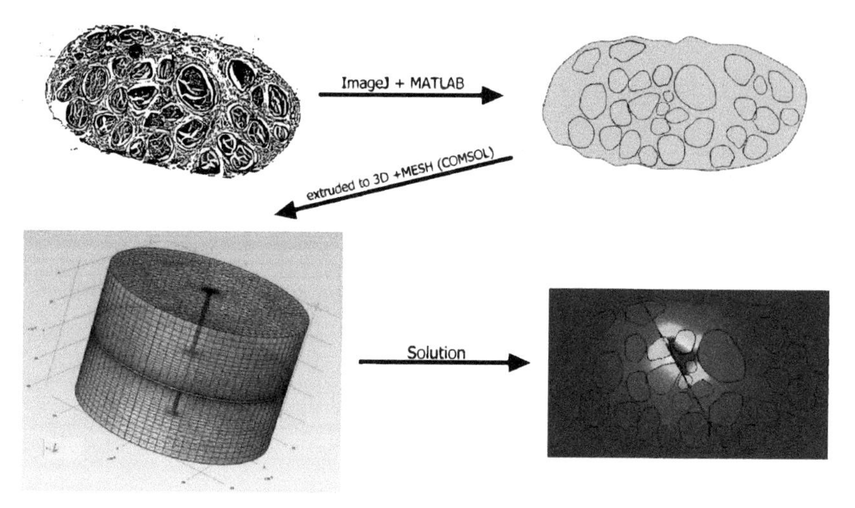

Figure 4.2 FEM solution. (a) Picture of cross-section of human median nerve. (b) 2D cross-section in COMSOL. (c) Final mesh of the entire structure in 3D. (d) Solution of the fem. Electric potential in plan xy (z = 0).

saline solution, but should be corrected in the future works about chronically implanted interfaces.

Models of electrodes are implemented separately and merged with the neural structure. Since the frequency range, which is of interest in sensing prosthetic applications, is low, we can assume a quasistatic approximation of Maxwell's equations within the nerve volume (Bossetti et al., 2008). Therefore, the electromagnetic problem can be expressed through Laplace formulation for the extracellular electric potential (Veltink et al., 1988; McIntyre and Grill, 2002):

$$\nabla \cdot \sigma \nabla Ve = 0 \tag{4.1}$$

To optimize the model from the computational load standpoint, an infinite-length/infinite-diameter to finite-length/finite-diameter approximation has to be considered. While in physics the 0-voltage is defined in infinity, within FEM model, in order to emulate the proper boundary conditions of the problem, the ground condition is set to the outermost surface of a finite model (McIntyre and Grill, 2002). Taking into account limited resources and time constraints, a minimal sufficient boundary dimensions had to be found calculating appropriated indexes (Raspopovic et al., 2011). Sufficient meaning the solution needs to be electromagnetically correct.

4.3 Neuron Fiber Model

To model the dynamics of nerve fiber, MRG (McIntyre Richardson Grill) model was used (McIntyre, 2002). This model represents the nonlinear modified Hodgkin–Huxley equations for the active compartment of the axons (the nodes of Ranvier) and a detailed realistic representation of the myelinated tracts. The success of this model is debt to its capability to reproduce several experimental aspects of cells dynamics, and to its availability: it can be found in model repository of NEURON. The difference between state-of-the-art models resides basically on two aspects: the first is the membrane dynamics and the second is the representation of the compartments. Membrane dynamics refers to differential equations of the membrane potential and extracellular potential relation (i.e., numbers of ion channels implemented). While compartment representations refer to the number and type of compartments. In this case, MRG model introduced Na+, K+, leakage channel, and nap channel for reproducing the hyperpolarization on the recovery cycle. The sensory axons population were simulated in NEURON 7.3 as implemented in (McIntyre, 2002). For a fiber of diameter D, a model made of 21 nodes of Ranvier with internodal spacing $L = 100$ D was built.

On the other side, it is unknown where are placed the fibers groups, which convey a specific sensation: either over the whole fascicle, or only within a very limited area of it. Fibers vary in diameter and position within the fascicle. To address this issue, multiple populations could be generated to account for fascicles' anatomical variability. In sensory human nerves, the probabilistic distribution for fibers diameter resulted in two Gaussian distributions, which differentiated nociceptive fibers from fibers responsible for pressure/touch sensation. Furthermore, nodal length was fixed at 1 um and nodal diameter was scaled from (McIntyre and Grill, 2002). A total amount of 100 fibers, were placed randomly in the specific target fascicle (Figure 4.3). Finally, we considered that fibers within a specific fascicle innervate the same portion of the hand (Jabaley et al., 1980).

4.4 Hybrid Model Solution

The FEM was solved with a stationary solver considering the quasistatic approximation for the electromagnetic problem, i.e., an electrostatic problem. The linear system obtained by FEM is symmetric and positive definite, therefore it can be solved by the Conjugate gradients method which applicable to sparse systems that are too large to be handled by a direct implementation.

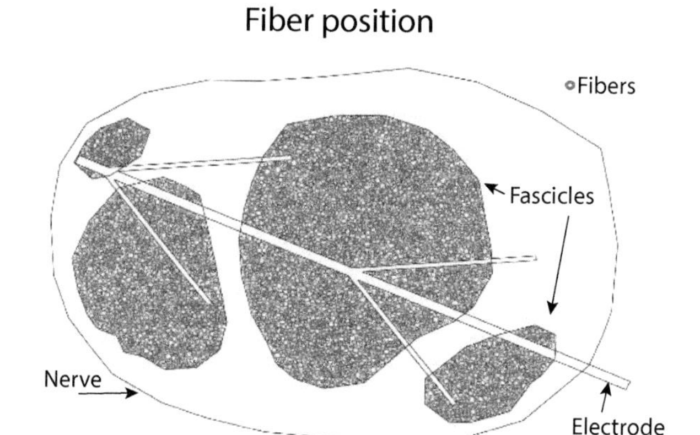

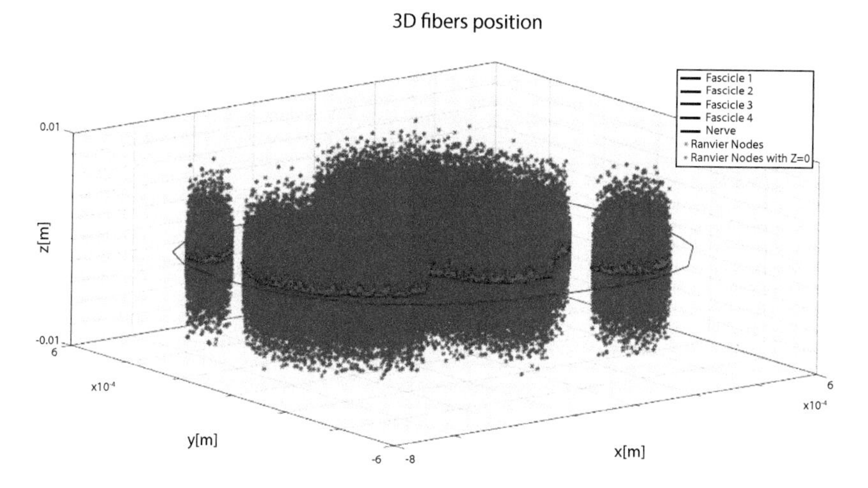

Figure 4.3 2D nerve cross-section with electrode and the fibers positioned inside the fascicles (red; left). 3D placement of Ranvier nodes for each fiber inside the nerve (right).

This method is an iterative solver that requires a preconditioner in order to improve its convergence. The preconditioner used was an algebraic multigrid, which is a numerical method that increases the computational speed by decreasing the complexity of the computations and then leading to a faster convergence. The convergence criterion is reached when the relative error becomes smaller than 1×10^{-6}.

Electric potentials generated inside the nerve by means of electrical stimulation were computed for the whole structure and then interpolated on the proper fibers positions and then neurons answer was computed

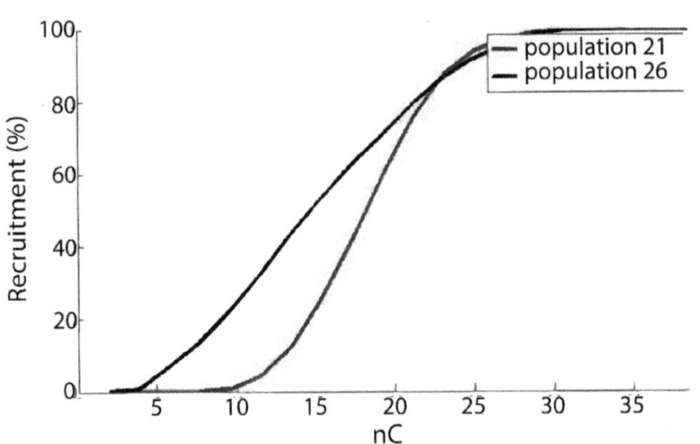

Figure 4.4 Active site of TIME close to three fascicles. (a) Recruitment curves (%). (b) Electric potential distribution (fixed 0–1.9 μV).

(Figure 4.4). Electric potentials were interpolated along the position of the nodes of Ranvier for each fiber in the model. Then, they were extracted from the FEM solutions and used as an extracellular mechanism for membrane depolarization. Fibers were stimulated by cathodal bipolar square current pulses of variable pulse-width (this is correct under quasistatic approximation). A fiber was considered recruited when a generated action potential traveled along its whole length (i.e., reached the last node of Ranvier).

4.5 Model-driven Electrode Design, Dimensions, and Number of Implants

The first, straightforward exploitation of models is for understanding of which type of electrode geometry is the most prominent for the selective stimulation of the discrete sensations. To do so, it is possible to implement different models of several electrodes type, which were successfully used in human applications, as intraneural and epineural electrode (Figure 4.5). Intraneural electrodes by design ensure closer distance to its targets then cuff-type electrodes. Results indicate that the most striking advantage of use of intraneural electrode is its one order of magnitude lower necessary charge threshold to elicit any fiber response respect to the case of epineural electrode (Raspopovic et al., 2017). On the other side, it is also possible to stimulate selectively the deep target fascicle by means of intraneural stimulation, while it is impossible to do so by means of epineural electrode. Finally, as regarding the dynamic of elicitable axonal response, it is possible to fine-modulate

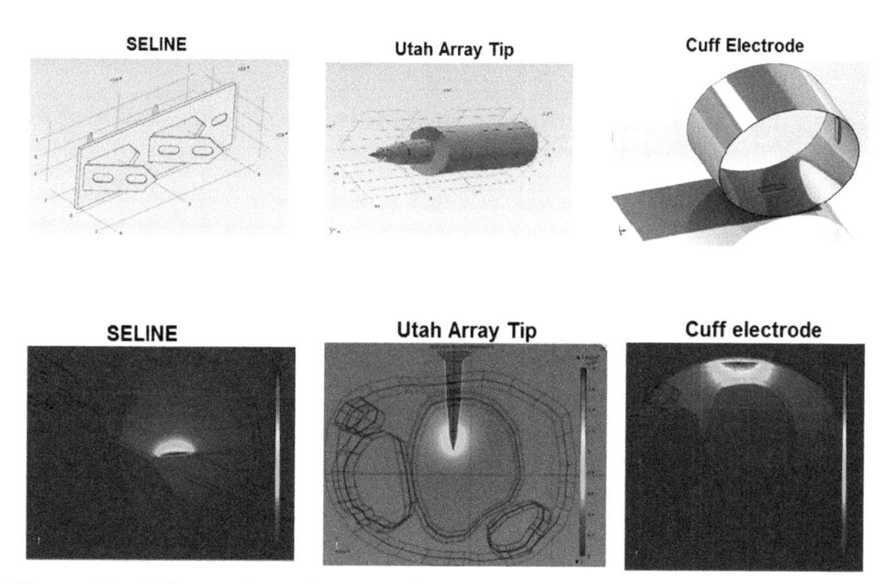

Figure 4.5 Different electrode geometries (top). FEM solutions according to different electrodes type (bottom).

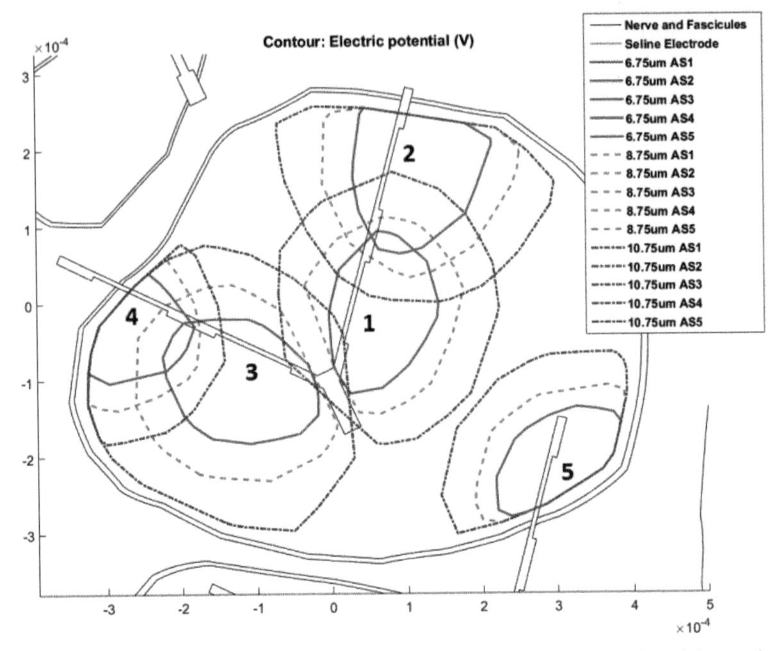

Figure 4.6 Isopotential curves regarding different fibers dimensions related for each active site inside the same fascicle (1–5).

the sensation by use of intraneural active site (TIME) and simple charge modulation, while the same is not feasible by epineural electrodes.

Appropriate electrode dimensions, number of active sites and their respective distances are essential for the manufacturing process. Models are ideal candidates for the proper addressing of this set of dimensions of electrodes. Total number of fascicles stimulated selectively is in a correlation with a number of contact sites, although some of them are recruited by more than one active site (Figure 4.6). The limiting factor, when having many active sites, is that the leads, necessary to connect them to the stimulator, are making the dimension of electrodes' substrate bigger, and therefore more invasive. The optimal number of electric active contacts for a neural electrode could be obtain using hybrid neuron model (Raspopovic et al., 2017).

In order to achieve the maximal performance (defined as the maximal possible number of fascicles elicited selectively) during the stimulation, with limited nerve damage, it is crucial to understand the optimal number of electrode for implant. This could be possible studying it with the models; indeed, it is possible to simulate different possible scenarios of implantation at the

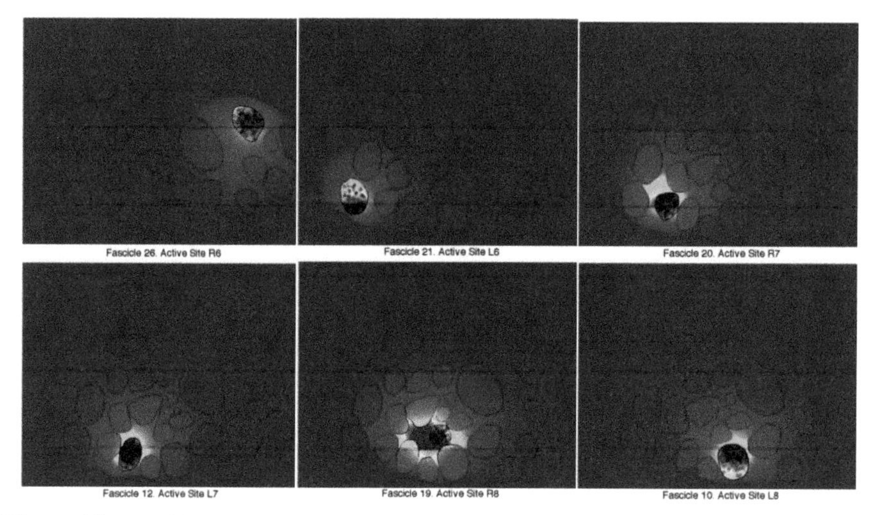

Figure 4.7 Double TIME implant in the same nerve and example of different stimulation positions.

same time (Figure 4.7). The most important goal is to reach the access to the maximal number of the fascicles using different active sites of the electrode. Technically, an implantation of many electrodes can be useful to stimulate every fascicle within the nerve, but too many electrodes could damage the nerve of the patient, put hard demand on the implantable electronics and transcutaneous communication implementation, and therefore this outcome of model is of essential value for neurosurgeon.

Using the hybrid computational model, it could be possible to design optimal configuration to stimulate the target nerve. Traditionally, the neural interface, even if having several stimulating contacts is generally used in paradigms concerning a single-channel: monopolar use. Monopolar stimulation consists in an activation of only one active site at time, while bipolar protocol enables to use two contacts in any configuration (with opposite or same polarity). This represents an example depicting the potentiality of the stimulation protocol design, guided by model findings.

4.6 Simulation of Biological Reaction to Electrode Optimization

Modeling framework is not only useful for the design of the neuroprostheses and their use. It can be successfully exploited also for the interpretation and investigation of scientific questions, which are not easy, or are impossible to face, by experimental empiric approach, that are of paramount importance for successful chronic use of neural interfaces.

One among the biggest issues with long-term use of neural interfaces in the neuroprosthetic devices is the decay/change of the stimulating capability over the time (Liu et al., 1999; Grill and Mortimer, 2000; Huang et al., 2004; McConnell et al., 2009; Polasek et al., 2009; Leach et al., 2010; Winslow

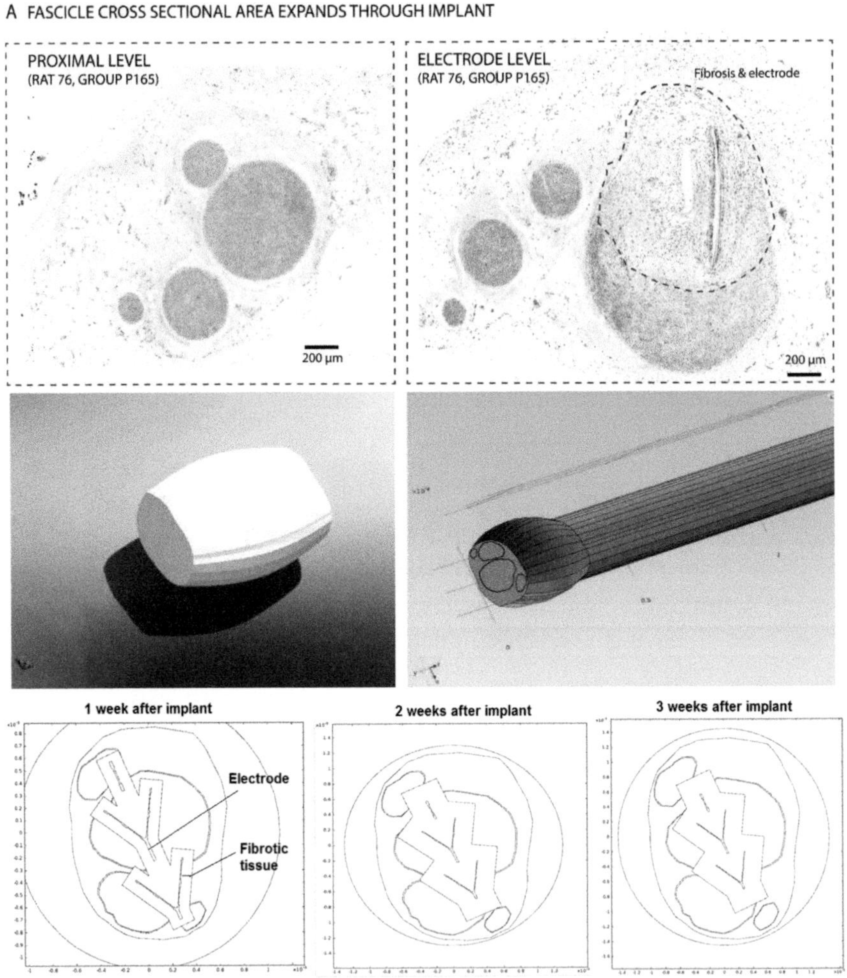

Figure 4.8 Modeling of the fibrotic tissue growth over weeks.

and Tresco, 2010; Raspopovic et al., 2014; Tan et al., 2014). Even though a significant amount of studies has been performed (Liu et al., 1999; Grill and Mortimer, 2000; Huang et al., 2004; McConnell et al., 2009; Polasek et al., 2009; Leach et al., 2010; Winslow and Tresco, 2010), the mechanisms of the thresholds change during the chronic neural stimulation are yet not elucidated. Between many possible hypothesis and interpretations, there is no consensus about the main factors; however, the nerve model can be an excellent instrument for testing some of these. It is possible to test several plausible hypotheses, which aim to explain the change of charge necessary to stimulate the nerve. For example: (i) Nerve fibers are in dysfunction/dying when electrode is placed intrafasciculary. (ii) Fibrotic tissue pushes electrode away from nerve fibers when electrode is placed intrafasciculary (Huang et al., 2004). (iii) Fibrotic tissue shifts electrode away from fascicle when placed extrafasciculary. (iv) Perturbation of the electric field by means of the fibrotic tissue generates the change in the axonal recruitment (Miocinovic et al., 2006). (v) Fibrotic encapsulation (Figure 4.8) changes over time the resistivity around the electrode.

4.7 Discussion

The hybrid modeling is a mandatory step in order to propose the optimized electrodes, and also to perform the most efficient manufacturing, avoid unnecessary animal experimentation, understand the unexpected changes and finally propose the hints for the neurosurgical procedure. It is of paramount importance to understand that, when dealing with models, they can be used properly only when addressing a clearly defined issue, and for what are tailored: it cannot be intended to explain all the aspects of such a complex system as neural system stimulation in every its aspect. Therefore, the models have to be customized toward the peculiar application of feature of interest. All of models account with specific limitations, and these should be clearly studied and stated, since could help in the correct interpretation of model results, their future exploitation and upgrading.

We believe that by the future development of the technologies, and specially imaging techniques, the sophisticated and widespread neuroprosthetic devices will go toward the ad hoc, ad personam modeling-based approach: starting from the high detailed images of structure of interest, and anatomical knowledge, by use of powerful computers, and efficient modeling computation we could have the patient-specific neural interface, and protocol of use.

References

Badia, J., Boretius, T., Andreu, D., Azevedo-Coste, C., Stieglitz, T. and Navarro, X. (2011). Comparative analysis of transverse intrafascicular multichannel, longitudinal intrafascicular and multipolar cuff electrodes for the selective stimulation of nerve fascicles, *Journal of Neural Engineering*, vol. 8, no. 3, p. 036023.

Boretius, T. et al. (2010). A transverse intrafascicular multichannel electrode (TIME) to interface with the peripheral nerve, *Biosensors and Bioelectronics*, vol. 26, no. 1, pp. 62–69.

Bossetti, C. A., Birdno, M. J. and Grill, W. M. (2008). Analysis of the quasi-static approximation for calculating potentials generated by neural stimulation, *J. Neural Eng.*, vol. 5, no. 1, pp. 44–53.

Capogrosso, M. et al. (2013). A computational model for epidural electrical stimulation of spinal sensorimotor circuits, *J. Neurosci.*, vol. 33, no. 49, pp. 19326–19340.

Coburn, B. and Sin, W. K. (1985). A theoretical study of epidural electrical stimulation of the spinal cord part I: Finite element analysis of stimulus fields, *IEEE Transactions on Biomedical Engineering*, vol. 11, no. BME-32, pp. 971–977.

Coburn, B. (1985). A theoretical study of epidural electrical stimulation of the spinal cord – Part II: Effects on long myelinated fibers, *IEEE Transactions on Biomedical Engineering*, vol. 11, no. BME-32, pp. 978–986.

Cutrone, A. et al. (2015). A three-dimensional self-opening intraneural peripheral interface (SELINE), *J. Neural Eng.*, vol. 12, no. 1, p. 016016.

Fang, Z. P. and Mortimer, J. T. (1991). Selective activation of small motor axons by quasi-trapezoidal current pulses, *IEEE Trans. Biomed. Eng.*, vol. 38, no. 2, pp. 168–174.

Frankemolle, A. M. M. et al. (2010). Reversing cognitive-motor impairments in Parkinson's disease patients using a computational modelling approach to deep brain stimulation programming, *Brain*, vol. 133, no. Pt 3, pp. 746–761.

Grill, W. M. and Mortimer, J. T. (1995). Stimulus waveforms for selective neural stimulation, *IEEE Engineering in Medicine and Biology Magazine*, vol. 14, no. 4, pp. 375–385.

Grill, W. M. and Mortimer, J. T. (1997). Inversion of the current-distance relationship by transient depolarization, *IEEE Transactions on Biomedical Engineering*, vol. 44, no. 1, pp. 1–9.

Grill, W. M. and Mortimer, J. T. (2000). Neural and connective tissue response to long-term implantation of multiple contact nerve cuff electrodes, *J. Biomed. Mater. Res.*, vol. 50, no. 2, pp. 215–226.

Grinberg, Y., Schiefer, M. A., Tyler, D. J. and Gustafson, K. J. (2008). Fascicular perineurium thickness, size, and position affect model predictions of neural excitation, *IEEE Transactions on Neural Systems and Rehabilitation Engineering*, vol. 16, no. 6, pp. 572–581.

Hennings, K., Arendt-Nielsen, L., Christensen, S. S. and Andersen, O. K. (2005). Selective activation of small-diameter motor fibres using exponentially rising waveforms: A theoretical study, *Medical & Biological Engineering & Computing*, vol. 43, no. 4, pp. 493–500.

Hines, M. L. and Carnevale, N. T. (1997). The NEURON simulation environment, *Neural Computation*, vol. 9, no. 6, pp. 1179–1209.

Huang, X., Nguyen, D., Greve, D. W. and Domach, M. M. (2004). Simulation of microelectrode impedance changes due to cell growth, *IEEE Sensors Journal*, vol. 4, no. 5, pp. 576–583.

Jabaley, M. E., Wallace, W. H. and Heckler, F. R. (1980). Internal topography of major nerves of the forearm and hand: A current view, *The Journal of Hand Surgery*, vol. 5, no. 1, pp. 1–18.

Leach, J. B., Achyuta, A. K. H. and Murthy, S. K. (2010). Bridging the divide between neuroprosthetic design, tissue engineering and neurobiology, *Front. Neuroeng.*, vol. 2, p. 18.

Liu, X., McCreery, D. B., Carter, R. R., Bullara, L. A., Yuen, T. G. and Agnew, W. F. (1999). Stability of the interface between neural tissue and chronically implanted intracortical microelectrodes. *IEEE Trans. Rehabil. Eng.*, vol. 7, no. 3, pp. 315–326.

McConnell, G. C., Rees, H. D., Levey, A. I., Gutekunst, C.-A., Gross, R. E., and Bellamkonda, R. V. (2009). Implanted neural electrodes cause chronic, local inflammation that is correlated with local neurodegeneration, *J. Neural. Eng.*, vol. 6, no. 5, p. 056003.

McIntyre, C. C. and Grill, W. M. (2002). Extracellular stimulation of central neurons: Influence of stimulus waveform and frequency on neuronal output, *Journal of Neurophysiology*, vol. 88, no. 4, pp. 1592–1604.

McIntyre, C. C., Richardson, A. G. and Grill, W. M. (2002). Modeling the excitability of mammalian nerve fibers: Influence of afterpotentials on the recovery cycle, *Journal of Neurophysiology*, vol. 87, no. 2, pp. 995–1006.

McNeal, D. R. (1976). Analysis of a model for excitation of myelinated nerve, *IEEE Transactions on Biomedical Engineering*, vol. BME-23, no. 4, pp. 329–337.

Miocinovic, S. et al. (2006). Computational analysis of subthalamic nucleus and lenticular fasciculus activation during therapeutic deep brain stimulation, *Journal of Neurophysiology*, vol. 96, no. 3, pp. 1569–1580.

Moffitt, M. A., McIntyre, C. C. and Grill, W. M. (2004). Prediction of myelinated nerve fiber stimulation thresholds: Limitations of linear models, *IEEE Trans. Biomed. Eng.*, vol. 51, no. 2, pp. 229–236.

Moraud, E. M. et al. (2016). Mechanisms underlying the neuromodulation of spinal circuits for correcting gait and balance deficits after spinal cord injury, *Neuron*, vol. 4, no. 89, pp. 814–828.

Navarro, X., Krueger, T. B., Lago, N., Micera, S., Stieglitz, T. and Dario, P. (2005). A critical review of interfaces with the peripheral nervous system for the control of neuroprostheses and hybrid bionic systems, *J. Peripher. Nerv. Syst.*, vol. 10, no. 3, pp. 229–258.

Oddo, C. M. et al. (2016). Intraneural stimulation elicits discrimination of textural features by artificial fingertip in intact and amputee humans, *Elife*, vol. 5, p. e09148.

Polasek, K. H., Hoyen, H. A., Keith, M. W., Kirsch, R. F. and Tyler, D. J. (2009). Stimulation stability and selectivity of chronically implanted multicontact nerve cuff electrodes in the human upper extremity, *IEEE Trans. Neural Syst. Rehabil. Eng.*, vol. 17, no. 5, pp. 428–437.

Raspopovic, S., Capogrosso, M. and Micera, S. (2011). A computational model for the stimulation of rat sciatic nerve using a transverse intrafascicular multichannel electrode. *IEEE Trans. Neural Syst. Rehabil. Eng.*, vol. 19, no. 4, pp. 333–344.

Raspopovic, S., Capogrosso, M., Badia, J., Navarro, X. and Micera, S. (2012). Experimental validation of a hybrid computational model for selective stimulation using transverse intrafascicular multichannel electrodes, *IEEE Transactions on Neural Systems and Rehabilitation Engineering*, vol. 20, no. 3, pp. 395–404.

Raspopovic, S. et al. (2014). Restoring natural sensory feedback in real-time bidirectional hand prostheses, *Science Translational Medicine*, vol. 6, no. 222, p. 222ra19.

Raspopovic, S., Petrini, F. M., Zelechowski, M. and Valle, G. (2017). Framework for the development of neuroprostheses: From basic understanding by sciatic and median nerves models to bionic legs and hands, *Proceedings of the IEEE*, vol. 105, no. 1, pp. 34–49.

Rattay, F. (1986). Analysis of models for external stimulation of axons, *IEEE Transactions on Biomedical Engineering*, vol. BME-33, no. 10, pp. 974–977.

Rattay, F. (1989). Analysis of models for extracellular fiber stimulation, *IEEE Transactions on Biomedical Engineering*, vol. 36, no. 7, pp. 676–682.

Rattay, F., Minassian, K. and Dimitrijevic, M. R. (2000). Epidural electrical stimulation of posterior structures of the human lumbosacral cord: 2. Quantitative analysis by computer modeling, *Spinal Cord*, vol. 38, no. 8, pp. 473–489.

Riso, R. R. (1999). Strategies for providing upper extremity amputees with tactile and hand position feedback – moving closer to the bionic arm, *Technol Health Care*, vol. 7, no. 6, pp. 401–409.

Saal, H. P. and Bensmaia, S. J., (2015). Biomimetic approaches to bionic touch through a peripheral nerve interface, *Neuropsychologia*, vol. 79, pp. 344–353.

Saal, H. P., Delhaye, B. P., Rayhaun, B. C. and Bensmaia, S. J. (2017). Simulating tactile signals from the whole hand with millisecond precision, *Proceedings of the National Academy of Sciences*, vol. 114, no. 28, pp. E5693–E5702.

Schiefer, M. A., Triolo, R. J. and Tyler, D. J. (2008). A model of selective activation of the femoral nerve with a flat interface nerve electrode for a lower extremity neuroprosthesis, *IEEE Transactions on Neural Systems and Rehabilitation Engineering*, vol. 16, no. 2, pp. 195–204.

Schiefer, M. A., Tyler, D. J. and Triolo, R. J. (2012). Probabilistic modeling of selective stimulation of the human sciatic nerve with a flat interface nerve electrode, *J. Comput. Neurosci.*, vol. 33, no. 1, pp. 179–190.

Schiefer, M., Tan, D., Sidek, S. M. and Tyler, D. J. (2016). Sensory feedback by peripheral nerve stimulation improves task performance in individuals with upper limb loss using a myoelectric prosthesis, *J. Neural. Eng.*, vol. 13, no. 1, p. 016001.

Tan, D. W., Schiefer, M. A., Keith, M. W., Anderson, J. R., Tyler, J. and Tyler, D. J. (2014). A neural interface provides long-term stable natural touch perception, *Sci. Transl. Med.*, vol. 6, no. 257, p. 257ra138.

Vuckovic, A., Rijkhoff, N., J. M. and Struijk, J. J. (2004). Different pulse shapes to obtain small fiber selective activation by anodal blocking – A simulation study, *IEEE Transactions on Biomedical Engineering*, vol. 51, no. 5, pp. 698–706.

Veltink, P. H., van Alste, J. A. and Boom, H. B. K. (1988). Simulation of intrafascicular and extraneural nerve stimulation, *IEEE Trans. Biomed. Eng.*, vol. 35, no. 1, pp. 69–75.

Winslow, B. D. and Tresco, P. A. (2010). Quantitative analysis of the tissue response to chronically implanted microwire electrodes in rat cortex, *Biomaterials*, vol. 31, no. 7, pp. 1558–1567, 2010.

Zierhofer, C. M. (2001). Analysis of a linear model for electrical stimulation of axons – Critical remarks on the "activating function concept", *IEEE Trans. Biomed. Eng.*, vol. 48, no. 2, pp. 173–184.

5

Biocompatibility of the TIME Implantable Nerve Electrode

Jordi Badia[1,2], Aritra Kundu[3], Kristian R. Harreby[3], Tim Boretius[4,5], Thomas Stieglitz[5,6,7], Winnie Jensen[3,*] and Xavier Navarro[1,2,*]

[1]Institute of Neurosciences and Department of Cell Biology, Physiology and Immunology, Universitat Autònoma de Barcelona, Bellaterra, Spain
[2]Centro de Investigación Biomédica en Red sobre Enfermedades Neurodegenerativas (CIBERNED), Spain
[3]Department of Health Science and Technology, Aalborg University, Denmark
[4]Neuroloop GmbH, Freiburg, Germany
[5]Laboratory for Biomedical Microsystems, Department of Microsystems Engineering-IMTEK, Albert-Ludwig-University of Freiburg, Freiburg, Germany
[6]BrainLinks-BrainTools, Albert-Ludwig-University of Freiburg, Freiburg, Germany
[7]Bernstein Center Freiburg, Albert-Ludwig-University of Freiburg, Freiburg, Germany
E-mail: xavier.navarro@uab.cat
*Corresponding Authors

5.1 Introduction

A key component for the clinical applicability of neuroprostheses is the neural electrode, intended to bidirectionally exchange information with the nervous system, thus allowing recording of nerve signals and stimulation of nerves and muscles over extended periods of time. A prerequisite for the application of nerve electrodes is that the implant must be biocompatible.

Biocompatibility can be defined as "the ability of a material to perform an appropriate host response in a specific application" (Williams, 1987). The compatibility between a technical and a biological system can be divided into *structural biocompatibility* and *surface biocompatibility* (Stieglitz, 2004). Structural biocompatibility refers to the adaptation of the artificial structure to the mechanical properties of the surrounding tissue, so that the device design and material properties should adapt to the biological structure of the target tissue. Surface biocompatibility deals with the interaction of chemical, physical, and biological properties of the foreign material and the target tissue. A material can be considered biocompatible if substances are only released in nontoxic concentrations and the biological environment reacts only with a mild foreign body reaction and encapsulation with connective tissue. The design and size, as well as the material choice and the interface surface, have to ensure stable properties of the electrode-electrolyte interface throughout the implant lifetime (Navarro et al., 2005). Once implanted, a neural interface has to remain within the body of the subject for months or years, so the stability of the materials in the electrode is crucial. The electrode has to be resistant to corrosion during stimulation and to the attack of biological fluids, enzymes, and macrophages produced during the initial foreign body reaction. It must be composed of inert materials, both passively and when subjected to electrical stimulation, since deterioration of the device may result in implant failure and the release of toxic products.

The first step to determine if any electrode material is biocompatible is, on one side, the in vitro study, bringing cell cultures in contact with the material and evaluating different parameters as morphological and ultrastructural changes (Koeneman et al., 2004; Vince et al., 2004), and, on another side, the in vivo subcutaneous implant of the material to investigate the tissue reaction, evaluating the thickness of cellular layers that surround the material and the presence of inflammatory cells (Vince et al., 2004). The following step is the evaluation of the full electrode device implanted in the target tissue, in this case the peripheral nerve, chronically, assessing possible functional changes and structural damage to the nerve as a consequence of the implanted device.

The transverse intrafascicular multichannel electrode (TIME) (Boretius et al., 2010) is intended to be implanted transversally in the peripheral nerve and address several subgroups of nerve fibers with a single device. Therefore, in an initial stage of the TIME project we performed extensive in vivo studies in order to assess the biocompatibility and safety of TIMEs after implantation in the rat and the pig.

5.2 Biocompatibility of the TIME in the Rat Nerve Model

5.2.1 Biocompatibility of the Substrate and Components

Following the standards defined in the ISO-10993 protocol for testing local effects after implantation of a device, pieces of polyimide substrate containing deposited iridium oxide (IrOx) as conductor for the active sites of the device were implanted in the subcutaneous tissue of adult rats during 4 weeks. The hexagonal pieces of polyimide substrate, with a surface of 100 mm^2, containing 19 deposited circles of IrOx (occupying a total of 50 mm^2) were provided by IMTEK. After shaving and disinfecting the rat skin, four incisions 1.5 cm long were done bilaterally on the back of the animals, with a distance of 2.5 cm between incisions. Sterile pieces of polyimide or of silicone as control, were implanted subcutaneously, one in each prepared subcutaneous pocket. The incisions were then sutured and disinfected. After 4 weeks, animals were sacrificed, and perfused transcardially with 4% paraformaldehyde. The implanted specimens were removed with the surrounding skin tissue and processed for histological and immunohistochemical analyses.

Compared to the intact skin, samples that had the polyimide implant, as well as silicone sheath implant as control, showed the epidermis and the connective tissue layer, with normal organization and appearance, as revealed by hematoxylin-eosin staining (Figure 5.1).

There were not areas of necrotic tissue in the subcutaneous and muscle layers. However, both silicone and polyimide implants were surrounded by a dense thin capsule of connective tissue. The thickness of the fibrous capsule was larger in silicone sections (outer zone: 159 ± 35 μm; inner zone: 69 ± 21 μm) than in polyimide sections (outer zone: 107 ± 9 μm; inner zone: 60 ± 7 μm). Immunohistochemical labeling was used to assess the inflammatory reaction by estimating the presence of macrophages stained with Iba-1 antibody in the capsule surrounding the implants. In samples with silicone implants macrophages were present at the inner zone, near blood vessels, but not at the outer and lateral zones. Sections with polyimide implants showed also concentration of macrophages in the inner zone, and lower amount at lateral zones.

Polyimide-based materials have been demonstrated biocompatible with respect to toxicity (Richardson et al., 1993; Rihová, 1996) as well as to biostable in other types of electrodes implanted in vivo (Rodríguez et al., 2000; Ceballos et al., 2003; Lago et al., 2007). The histological biocompatibility assessment of the tissue surrounding the implants of polyimide

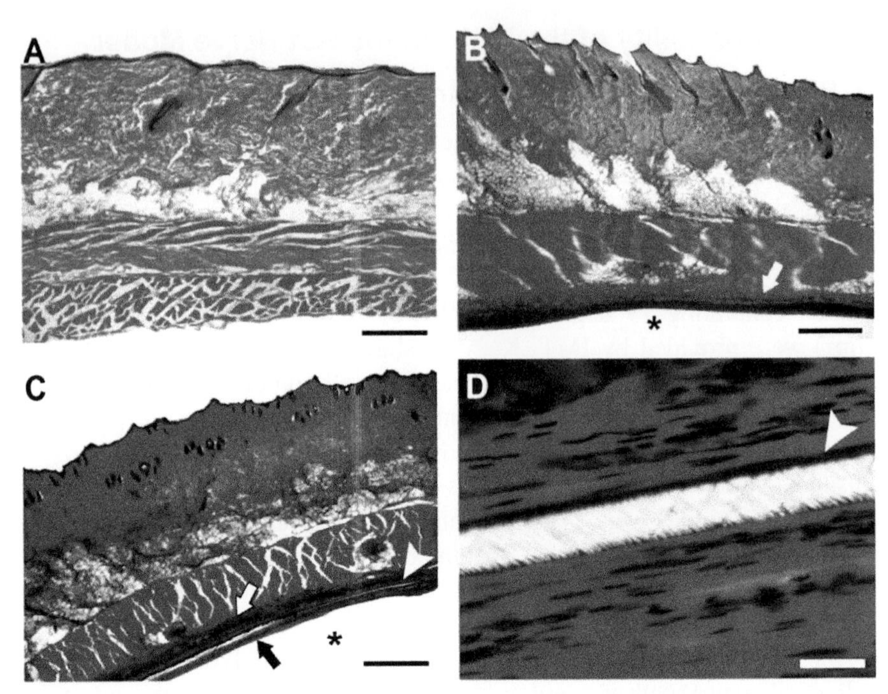

Figure 5.1 Hematoxylin-eosin staining of representative sections from skin without implant (A), with a silicone implant (B) and with a polyimide implant ((C), and detail in (D)). (B) Silicone implants showed a cavity formed during tissue processing (asterisk). White arrows show the superficial fibrous layer (outer zone). (C) Polyimide implants were embedded in the fibrous capsule (white arrowheads). The deep fibrous layers of the capsule are also shown (inner zone, black arrow). Scale bars: 500 μm ((A)–(C)) and 20 μm (D).

containing IrOx dots, selected to fabricate the TIME, indicated good integration in the skin and subcutaneous tissue of the adult rat. This is in agreement with previous studies, describing the absence of a gross response to metal-coated pieces of polyimide or uncoated control silicone (Cogan et al., 2003). Although the capsule surrounding the implant was observed around both the unprocessed silicone and the metalized polyimide samples, it was significantly thinner in the latter. Macrophages were slightly more abundant next to polyimide than to silicone samples, but still their density was low, suggesting a slight degree of inflammatory response around the implanted materials.

5.2.2 Biocompatibility of the TIME Implanted in the Rat Nerve

The electrodes used in the study corresponded to the TIME-2 and TIME-3 designs, explained in the previous chapter, and were produced by the Department of Microsystems Engineering (IMTEK) of the University of Freiburg. Cleaned and sterilized TIMEs were implanted into the sciatic nerve of adult rats. In one group (Acute, $n = 7$) of rats the TIME-2 was implanted and, in order to evaluate the damage just induced by the surgical implantation procedure, the electrode was retired after an acute electrical stimulation protocol was performed. In two other groups (TIME-2 and TIME-3, $n = 5$ each) a TIME-2 or a TIME-3 were transversally implanted in the sciatic nerve and remained for 2 months.

The TIME devices were transversally inserted across the three fascicles of the sciatic nerve (sural, tibial, and peroneal branches) at the midthigh. The electrode was inserted with the help of a small straight needle attached to a 10-0 loop thread that was passed between the two arms of the TIME. The needle was inserted transversally across the sciatic nerve and then pulled the thin-film structure through it. Once the structure was implanted, the needle was removed by cutting the suture. The TIME-2 ribbon was routed across the overlying muscular plane to the lateral thigh region, where the ending part was secured under the skin. For the TIME-3 design, the ribbon was routed along the nerve and the ending pad was accommodated under the muscle (Figure 5.2).

All the animals were followed up with a battery of neurophysiological tests to obtain evidence of possible functional alterations, and final histological analysis to assess potential damage induced by the implanted TIMEs on the nerve. The study design aimed to discriminate effects due to the implantation procedure, to the presence of the intraneural electrode segment, and to the mechanical motion induced by the TIME ribbon and connector during chronic implants (Badia et al., 2011).

During the 2 months implantation time, there were no remarkable changes in any of the parameters of motor and sensory nerve conduction tests performed in any of the implanted groups, Acute, TIME-2, and TIME-3. The amplitude of the compound action potentials (CMAPs and CNAPs) obtained by stimulation of the right sciatic nerve after TIME-2 and TIME-3 implantation did not change notably at any interval postimplantation in comparison with the control nerve values. There was evidence of slight slowing of nerve conduction velocity at 7 days in the implanted groups, but it was normalized

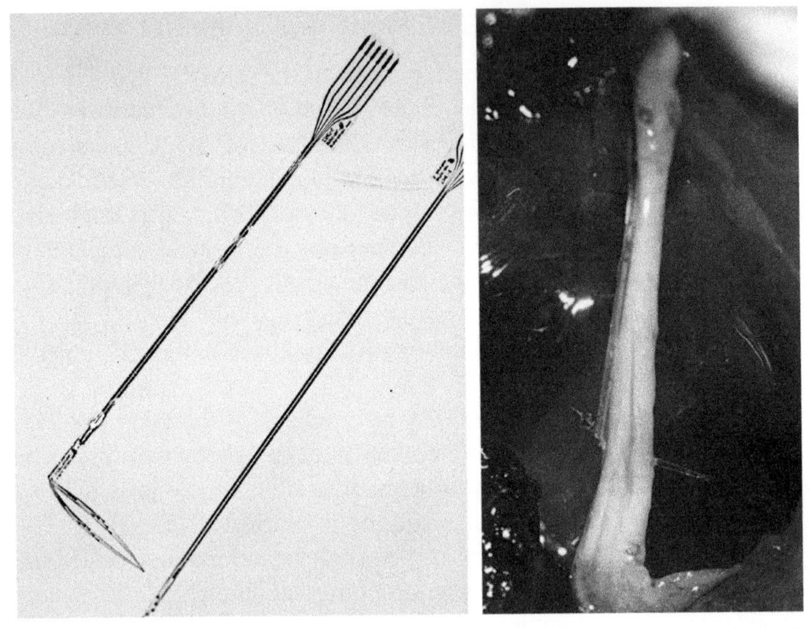

Figure 5.2 Left: Photograph of a TIME-3 in which the intraneural portion is angled at 90°. Right: photograph of the insertion of a TIME-3 with the ribbon accommodated to the longitudinal axis of the sciatic nerve.

at 30 days. The implanted nerves showed similar values over time than the contralateral intact nerves (Figure 5.3A, B).

Walking track test, used to evaluate the locomotion performance, did not show variations between the right (TIME-2 and TIME-3 implanted) and left (intact) hindpaw prints along the follow-up. The Sciatic Functional Index averaged between −5 and +20 during follow-up (Figure 5.3C). The algesimetry tests yielded similar values of the pain threshold for withdrawal between the implanted and the contralateral sides, without evidence of hyperalgesia that might be induced by nerve compression or injury (Figure 5.3D). No loss of pain sensitivity was appreciated in any area of the hindpaw under pinprick testing.

5.2.3 Morphological Evaluation of the Implanted Nerves

The macroscopic examination during final dissection at 2 months after implantation showed that the electrodes remained in place within the sciatic

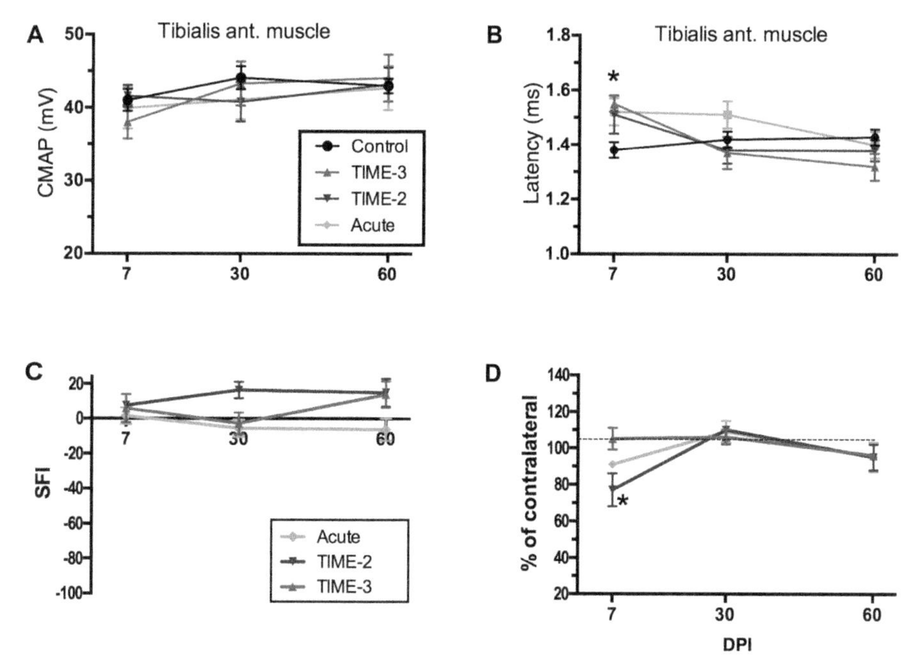

Figure 5.3 Neurophysiological tests result in the three groups with Acute, TIME-2, and TIME-3 chronic implant in comparison with control values. Values of the CMAP amplitude (A) and onset latency (B) of the tibialis anterior muscle. Values of the SFI (C) and of algesimetry (D). Modified from Badia et al., 2011.

nerve in all animals of groups TIME-2 and TIME-3. Transverse sections at the segment that contained the intraneural TIME allowed to see the electrode strip crossing the fascicles of the sciatic nerve, covered by a fibrous tissue that disrupted the microarchitecture of the nerve (Figure 5.4).

Transverse semithin sections of the nerve distal to the insertion site showed a normal fascicular organization (Figure 5.4). The numbers of myelinated fibers counted in the distal level for each of the three branches of the sciatic nerve were in all the implanted groups similar to those found in control intact nerves. The reduction of about a 10% in the mean number of myelinated fibers of the sciatic nerve in group TIME-2 was mostly due to a decrease of myelinated fibers of about the 8% in the tibial nerve and 25% in the sural nerve, the smallest of the three branches. Group TIME-3 has only a small loss of myelinated fibers in the sural nerve (about 20%) (Badia et al., 2011).

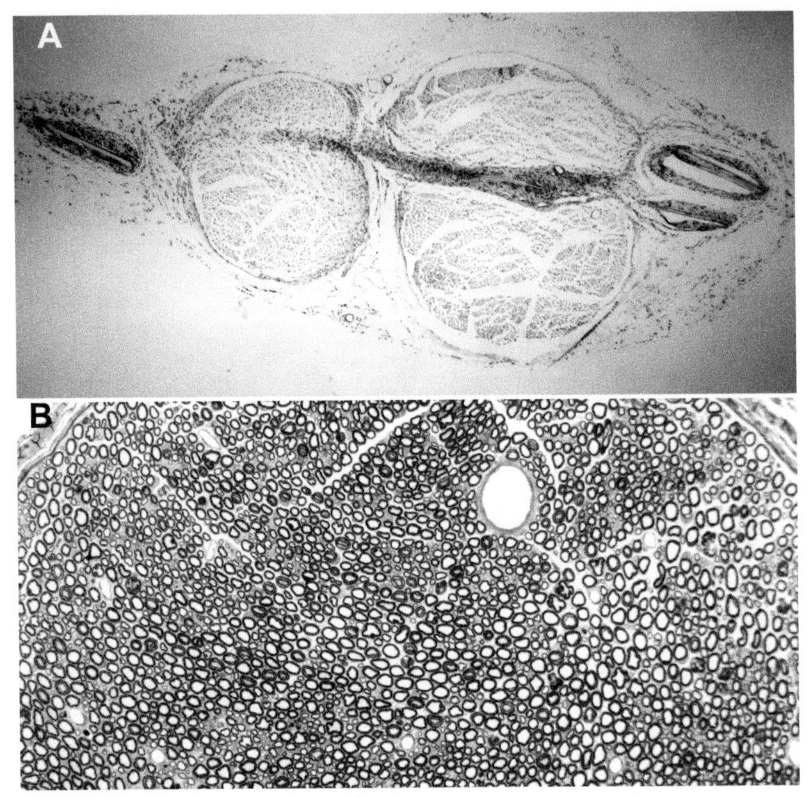

Figure 5.4 Cross-sections of an implanted rat sciatic nerve. (A) At the level of TIME-3 implant crossing the tibial branch and part of the peroneal branch. Note the mild fibrous tissue surrounding the electrode. (B) Semithin transverse section of the tibial nerve of the same animal distal to the implant site. There are no signs of degeneration and the density of myelinated fibers is similar to controls.

5.3 Biocompatibility of the TIME in the Pig Nerve Model

Device implantation and associated tissue injury trigger a cascade of inflammatory and wound healing responses that are typical of a foreign body response (FBR) (Morai et al., 2010). The continuous presence of biomaterial devices leads to chronic inflammation. The wound healing response depends on the size and extent of the implant. The end stage of FBR involves shielding the implant by a vascular and collagenous fibrous capsule (Williams et al., 1983; Labat-Robert, 1990; Kovacs et al., 1991). The response to any implant is wound healing comprised of hemostasis (seconds to hours), inflammation (hours to days), repair (days to weeks), and remodeling (weeks to months)

(Stroncek, 2008). To evaluate the biological effect of the TIME electrodes in the large nerve animal model (chronically implanted Göttingen minipigs), and particularly the fibrotic capsule formed around the electrode, we harvested and analyzed nerves following several weeks of implant.

All experimental procedures were approved by the Animal Experiments Inspectorate under the Danish Ministry of Justice. Female, Göttingen minipigs. Under full anesthesia and using aseptic surgical techniques one or two TIMEs were implanted in the median nerve above the elbow joint. If only one TIME electrode was implanted, it was placed at an angle of 90° or 135°. When two electrodes were implanted they were implanted with a 45° difference in angle. Lead-out wires were tunneled subcutaneously to the back of the pig where they excited the skin.

To collect tissue for histological evaluation an incision was made over the implant area and the location of the TIME and the median nerve were identified. The animal was then euthanized and the nerve including electrode(s) and lead-out wires was harvested. We performed further dissection of the harvested tissue in order to free the nerve and electrode from the surrounding fibrotic tissue (Figure 5.5). A specimen of approx. 5 mm of the nerve containing the TIME electrode was taken and either immersed in formalin or in liquid nitrogen. To maintain orientation of the nerve the proximal cross-section was marked with green dye while the entry and the exit point of the identified TIME electrodes were marked with blue. In one animal, we collected control samples from the right, nonimplanted leg. We included 11 animals in the study that were implanted during 22 ± 7.7 days (range 8–37 days).

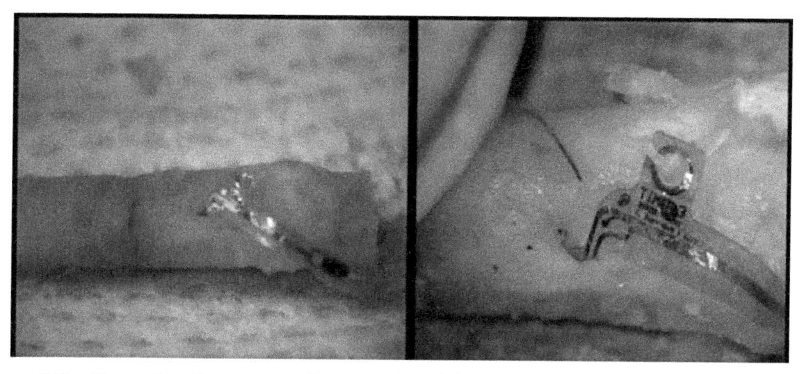

Figure 5.5 Example of nerve specimen retrieved from a minipig after approx. 30 days of implant. The surrounding fibrotic tissue has been removed by careful dissection to identify the entry and exit points of the TIME.

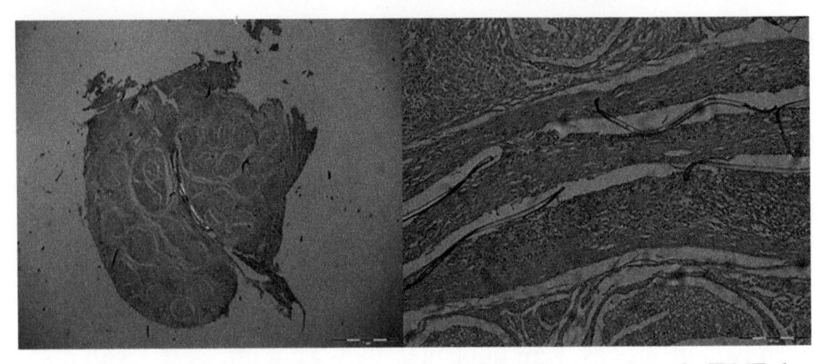

Figure 5.6 Typical samples of H&E stains of the peripheral nerve, where the TIME electrode has been identified inside the nerve. Left (×20, Pig 02): whole nerve with TIME transversing through the nerve easily identified. Right (×100, Pig 02): higher magnification of the implant site – the TIME electrode and a layer of fibrosis surrounding the electrode is seen. The visible "cracks" inside the fascicles result from the processing and embedding the nerve.

Transverse sections of 5 μm thickness were cut in the nerve using a cryostat. The tissue was stained with Hematoxylin and Eosin (H&E) (Figure 5.6). To estimate the thickness of the fibrotic scarring, digital pictures were taken through a microscope. To avoid bias we randomly chose five points and measured the distance from the polyimide structure of the electrode to the rim of the fibrotic capsule perpendicularly.

5.3.1 Morphological Evaluation of the Implanted Nerves

All animals were in good health state and supported weight on the implanted leg. In approximately half of the animals we noticed signs of swelling, edema, and tenderness at the wound following the implant, which disappeared in the following 6–9 days. Eight of the 13 animals developed infection either at the implant wound or at the percutaneous connector in the back. In the majority of the cases we were able to treat the infection to a point where it was not possible to detect it visually.

Layer of fibrosis. Complete visualization of the entire length of TIMEs placed transversely through the nerves was not always possible. The estimated layer of fibrosis is shown in Table 5.1. The thickness of the fibrosis averaged 108 ± 40 μm. A similar FBR was also identified in the histological evaluation of the rat sciatic nerve. We found no apparent correlation between the thickness of the fibrosis and the duration of the implant.

Table 5.1 Estimated thickness of the fibrotic capsule formed around nine TIME electrodes implanted in seven pigs

Animal	Pig 01	Pig 02	Pig 04	Pig 06	Pig 09	Pig 10 TIME-1	Pig 10 TIME-2	Pig 11 TIME-1	Pig 11 TIME-2
Thickness [μm] Mean ± SD	74.3 ± 8.7	62.0 ± 26.5	179.6 ± 60.6	71.4 ± 23.5	128.3 ± 42.9	116.4 ± 9.4	141.6 ± 30.6	117.9 ± 38.9	100.4 ± 21.1
# days implant	36	19	25	20	8	31	31	37	37

Macroscopic changes. For animals included in the "Chronic selectivity" experiments we found that typical inflammatory cells were present around the implant (i.e., including lymphocytes, macrophages, and giant cells), however some variation between animals was observed. Also we identified fibrocytes/fibroblasts in all animals. There were no signs of necrosis. The presence of inflammatory cells and fibroblasts/fibrocytes indicated that the wound healing process was likely still ongoing at the time of explant. It is important to note that the inflammatory response and the formation was only found around the electrode, i.e., the remaining part of the nerves and fascicles appeared normal and without presence of fibrosis or inflammatory response.

5.4 Discussion

Intraneural electrodes are intended to provide a good degree of sensitivity and selectivity for stimulation and recording action potentials at the nerve fibers of the implanted nerve fascicles. However, they have the risk of inducing damage to the nerve. Relative motion between the nerve and surrounding muscles during limb movement can exert forces on the electrode, and eventually extract it or damage the nerve if the electrode is unable to move with the nerve. Lead management and connector requirements offer additional challenges; lead wires or ribon strips can produce tethering forces on the electrode, resulting in damage to the nerve or breakage of the device. The materials used for the TIME, i.e., polyimide as substrate and IrOx as conductive sites, are biocompatible, as corroborated in our study in subcutaneous implants. The induced FBR was similar or slightly thinner than with a control substrate as silicone.

In general, we found that all the animals implanted, either rats or pigs, showed good recovery from the surgery and were in good health during the implant period. The infections experienced are likely related to the pig model, since no infections were observed in the rat model. However, the risk of infection in future human clinical experiments will be very low, since it is much easier to keep the percutaneous lead wires and exit points clean and protected.

From the surgical point of view, TIMEs were easy to implant in the nerve, because of the high flexibility and small thickness of the electrode strips, even in the small size nerves of the rat. The results of the follow-up evaluation indicate that either acute or chronic implantation of the TIME in the rat sciatic nerve for up to 2 months did not cause significant signs of axonopathy, axonal loss or demyelination, as evidenced by the functional and histological findings. Nerve conduction tests showed a mild increase in latency time at the first week after implantation, which is likely attributable to the surgical implantation procedure, since damage to the perineurium leads to endoneurial edema. Nevertheless, the degree of dysfunction was low and time limited, since the animals showed recovery toward normal values during the following weeks. Moreover, the nociceptive responses quantified by algesimetry tests showed a mild decrease of the threshold only at 7 days in the group TIME-2, whereas at 1 and 2 months there were no signs of hyperalgesia or pain. The absence of differences between the animals implanted with TIME-3 and their controls indicates that the tethering and motion forces, produced by the surrounding muscles and transmitted through the ribbon in the case of the TIME-2 implants, were minimized with the next TIME-3 design.

Regarding the histological results in the rat implanted nerves, the finding of a slightly reduced number of myelinated fibers distal to the implant site in group TIME-2 is suggestive of damage and subsequent axonal degeneration of a small population of nerve fibers. Nevertheless, we did not find images of ongoing degenerating fibers at 2 months, thus pointing to a time-limited damage, most likely occurring during insertion. Comparatively, the reduction of about a 10% in the number of myelinated nerve fibers with TIME-2 in the chronic group is similar to that previously reported after the implantation of other intrafascicular electrodes such as the polyimide thin film LIFE (Lago et al., 2007), which is coherent with the fact that the Young's modulus of the substrate material (polyimide, UBE U-Varnish-S) is similar between TIME and LIFE. The Utah Slanted Electrode Array (USEA) that is also inserted transversally to the nerve yielded results similar to the ones found for the TIME, although we did not find images of ongoing degenerative fibers at 2 months postimplant, in contrast with the results for the USEA (Branner et al., 2004); the difference may be attributed to the stiffness of the silicon USEA structure.

The reduction in the number of myelinated nerve fibers in the tibial nerve of the group TIME-2 was not observed in the group TIME-3. This difference can be explained because of the absence of traction forces with the TIME-3 design, this conclusion is reinforced by the fact that the nerves of the Acute group did not show fiber loss. The minor loss of myelinated fibers in group TIME-2 was not extensive enough to affect the animal function or electrophysiological responses. Since in group TIME-3 we did not find significant histological abnormalities, we can attribute most of the axonal loss in group TIME-2 to the effects produced by the traction forces for some days after implantation during the animal motion, and not to the material itself. The refinement in the geometrical design of the TIME-3 allows for a better adaptation of the structure to the anatomical properties of the nerve and to reduce the tethering forces to which the electrode may be subjected.

These conclusions were proved by the subsequent study of TIME-3 implants in the pig median nerve. There were no clinical evidences of nerve damage or dysfunction during several weeks of implant. The histological study of the harvested nerves did not show any appraten abnormality on the microstructure of the nerve, except for the presence of the transversal electrode, which was covered by a fibrous capsule, as corresponds to a normal FBR. The overall mean of the fibrosis was $108 \pm 40\ \mu m$, which is comparable to what is found in the literature (Williams et al., 1983; Labat-Robert, 1990; Kovacs et al., 1991). A similar reaction and layer of fibrosis was also identified in the histological evaluation of the rat sciatic nerve. Interestingly, there was no apparent correlation between the thickness of the fibrotic scarring and the duration of the implant, thus suggesting that the FBR occurs during the first 1–2 weeks after implant, and thereafter remains stable, helping also to maintain the implanted electrode in place. The inflammatory response and fibrosis was only found around the electrode, i.e., the remaining part of the nerves appeared normal.

Altogether our results indicated that the TIMEs are biocompatible and safe after chronic implantation even in a small peripheral nerve, such as the rat sciatic nerve. The mild effects of the TIME on the nerve will be minimized when implanting the electrode of the same dimensions in larger nerves of humans.

References

Badia, J., Boretius, T., Pascual-Font, A., Udina, E., Stieglitz, T. and Navarro, X. (2011). Biocompatibility of chronically implanted transverse intrafascicular multichannel electrode (TIME) in the rat sciatic nerve. IEEE Trans Biomed Eng 58:2324–2332.

Branner, A., Stein, R. B., Fernandez, E., Aoyagi, Y. and Normann, R. A. (2004). Long-term stimulation and recording with a penetrating microelectrode array in cat sciatic nerve. IEEE Trans Biomed Eng 51:146–157.

Ceballos, D., Valero-Cabré, A., Valderrama, E., Schüttler, M., Stieglitz, T. and Navarro, X. (2002). Morphological and functional evaluation of peripheral nerve fibers regenerated through polyimide sieve electrodes over long term implantation. J Biomed Mat Res 60:517–528.

Cogan, S. F., Edell, D. J., Guzelian, A. A., Liu, Y. P. and Edell, R. (2003). Plasma-enhanced chemical vapor deposited silicon carbide as an implantable dielectric coating. J Biomed Mat Res 67A:856–867.

ISO 10993-6:2007 (E). International Standard. 'Biological evaluation of medical devices. Part 6: Tests for local effects after implantation', 2007.

Koeneman, B. A., Lee, K. K., Singh, A., He, J., Raupp, G. B., Panitch, A. and Capco, D. G. (2004). An ex vivo method for evaluating the biocompatibility of neural electrodes in rat brain slice cultures. J Neurosci Methods 137:257–263.

Kovacs, E. J. (1991). Fibrogenic cytokines: The role of immune mediators in the development of scar tissue. Immunol Today 12:17–23.

Labat-Robert, J., Bihari-Varga, M. and Robert, L. (1990). Extracellular matrix. FEBS Lett 268:386–393.

Lago, N., Yoshida, K., Koch, K. P. and Navarro, X. (2007). Assessment of biocompatibility of chronically implanted polyimide and platinum intrafascicular electrodes. IEEE Trans Biomed Eng 54:281–290.

Morai, J., Papadimitrakopoulos, F. and Burgess, D. J. (2010). Biomaterials/tissue interactions: Possible solutions to overcome foreign body response. Am Assoc Pharmaceut Scien J 12:188–196.

Navarro, X., Krueger, T., Lago, N., Micera, S., Stieglitz, T. and Dario, P. (2005). A critical review of interfaces with the peripheral nervous system for the control of neuroprostheses and hybrid bionic systems. J Peripher Nerv System 10:229–258.

Richardson, R. R., Miller, J. A. and Reichert, W. M. (1993). Polyimides as biomaterials: Preliminary biocompatibility testing. Biomaterials 14:627–635.

Rihová, B. (1996). Biocmpatibility of biomaterials: Hemocompatibility, immunocompatibility and biocompatibility of solid polymeric materials and soluble targetable polymeric carrier. Adv Drug Deliv Rev 21:157–176.

Rodríguez, F. J., Ceballos, D., Schüttler, M., Valderrama, E., Stieglitz, T. and Navarro, X. (2000). Polyimide cuff electrodes for peripheral nerve stimulation. J Neurosci Methods 98:105–118.

Stieglitz, T. (2004). Considerations on surface and structural biocompatibility as prerequisite for long-term stability of neural prostheses. J Nanosci Nanotechnol 4:496–503.

Stroncek, J. D. and Monty Reichert W. (2008). Overview of wound healing in different tissue types. In 'Indwelling Neural Implants: Strategies for Contending with the *In Vivo* Environment', Reichert WM, editor. Boca Raton (FL): CRC Press, 2008.

Vince, V., Thil, M. A., Veraart, C., Colin, I. M. and Delbeke, J. (2004). Biocompatibility of platinum-metallized silicone rubber: In vivo and in vitro evaluation. J Biomater Sci Polym 15:173–188.

Williams, D. F. (1987). Definitions in biomaterials. *Progress in Biomedical Engineering* 4: 54. Elsevier Science: Amsterdam.

Williams, G. T. and Williams, W. J. (1983). Granulomatous inflammation – A review. J Clin Pathol 36:723–773.

6

Selectivity of the TIME Implantable Nerve Electrode

Jordi Badia[1,2], Kristian R. Harreby[3], Aritra Kundu[3], Tim Boretius[4,5], Thomas Stieglitz[5,6,7], Winnie Jensen[3,*] and Xavier Navarro[1,2,*]

[1]Institute of Neurosciences and Department of Cell Biology, Physiology and Immunology, Universitat Autònoma de Barcelona, Bellaterra, Spain
[2]Centro de Investigación Biomédica en Red sobre Enfermedades Neurodegenerativas (CIBERNED), Spain
[3]Department of Health Science and Technology, Aalborg University, Denmark
[4]Neuroloop GmbH, Freiburg, Germany
[5]Laboratory for Biomedical Microsystems, Department of Microsystems Engineering-IMTEK, Albert-Ludwig-University of Freiburg, Freiburg, Germany
[6]BrainLinks-BrainTools, Albert-Ludwig-University of Freiburg, Freiburg, Germany
[7]Bernstein Center Freiburg, Albert-Ludwig-University of Freiburg, Freiburg, Germany
E-mail: xavier.navarro@uab.cat
*Corresponding Authors

6.1 Introduction

For restoring the lost function of organs innervated by damaged peripheral nerves or to control a prosthesis substituting a lost organ, it is necessary to provide an array of electrodes with good capabilities for both stimulation and recording neural activity. Ideally, a bidirectional interface for the control of a bionic prostheses in amputees should allow on the one hand, recording of neural efferent motor signals to be used for the motion control of the mechanical prosthesis, and on the other, stimulating afferent sensory nerve fibers within the residual limb to provide sensory feedback to the user from tactile and force sensors embedded in the prosthesis. In amputees, the nerves in the stump, which previously innervated the missing limb, are still functional, even years after the amputation has occurred (Dhillon et al., 2005). If these nerves could be selectively interfaced, it would be possible to translate the motor signals into adequate actions and to provide the amputated subject with sensory input, perceived as originating from the missing limb.

The selection of a suitable nerve electrode has to consider a balance between invasiveness and selectivity, with the ultimate goal of achieving the highest selectivity for a high number of nerve fascicles by the least invasiveness and potential damage to the nerve. For controlling an advanced neuroprosthesis, multiple functions, either motor or sensory, conveyed by separate fascicles or bundles of axons in a peripheral nerve have to be interfaced. To adequately reproduce motor and sensory functions, it is, however, necessary that the interface achieves two conditions: topographical selectivity and functional selectivity. The knowledge of the fascicular topographical pattern of peripheral nerves of interest (Gustafson et al., 2009; Badia et al., 2010; Delgado-Martinez et al., 2016) will improve the implantation of electrodes in adequate positions, by placing at least one active site within each fascicle of interest. Regarding functional selectivity, for example, to induce a particular sensory input, e.g., touch sensation from the thumb, it is needed to selectively activate the specific nerve fibers that before amputation mediated touch information from the thumb to the brain. This should be done without recruiting fibers mediating information related to other sensory modalities or limb areas, which might overshadow the intended sensory input. The more individual parts of the nerve that can be selectively interfaced without recruiting other parts, the higher chance is that a specific part of the nerve can be targeted in a functional relevant way and the more selective the interface is. An intrinsic drawback, however, is that the invasiveness and the risk of the surgical implantation procedure tend to increase the more contact sites are placed in or around a nerve and the closer they are placed to individual nerve fibers.

The design of the transversal intrafascicular multichannel electrode (TIME) aimed to maximize the number of contact sites and their proximity to different populations of nerve fibers, while keeping the risk and invasiveness of the implantation low. The TIME is intended to be implanted transversally in the nerve and cross several fascicles or subgroups of nerve fibers with one single device (Boretius et al., 2010). By placing a number of active sites distributed along the intraneural implant, a single TIME is thus able to interface several fascicles, reducing the number of implanted devices with respect to other intrafascicular electrodes.

The TIME was extensively tested in computer simulations and animal studies. These studies had several purposes: (1) to develop and test surgical techniques, (2) to ensure the durability and biocompatibility of the electrode when exposed to the biological environment (see Chapter 5), (3) to evaluate the selectivity performance for stimulation and recording neural activity, and (4) to evaluate the performance over time. The current chapter focuses on the two last objectives.

6.2 Evaluation of TIME in the Rat Sciatic Nerve Model

6.2.1 Stimulation Selectivity

The testing in preclinical animal models is obliged before translation into human subjects to avoid any risk, and also to obtain adequate information regarding implantation and performance of the electrode. However, one disadvantage is that it is not possible to get a subjective description of the percepts induced by activating the nerve in these models, and thus the selectivity performance of the implanted electrode must be quantified by neurophysiological methods. In the following studies, for practicality, the tests for selectivity were performed in an inverse way to what it is pretended in human amputees, i.e., selectivity of stimulation of motor nerve fibers, and selectivity of recording of sensory nerve fibers.

6.2.1.1 Methods

Acute experiments were made on adult rats in which the sciatic nerve was implanted with an electrode for assessing selective electrical stimulation at the fascicular and subfascicular levels. We compared the results obtained with three types of neural electrodes: TIME, LIFE, and cuff (Badia et al., 2011). The three types of electrodes had adequate dimensions for the rat sciatic nerve and were made on polyimide substrate with platinum active sites and contact lines. In one group of rats (group TIME-A) the TIME (Boretius et al., 2010) was transversally implanted traversing the tibial and peroneal fascicles of the sciatic nerve. With the aim of assessing the selectivity of TIME just in the tibial nerve (subfascicular selectivity) the TIME was implanted in dorsoventral direction across the tibial nerve (group TIME-B). To compare the results of TIME with Cuff and tf-LIFE devices, cuff electrodes of 12 poles (four tripoles) (Navarro et al., 2001) were implanted around the sciatic nerve (group Cuff), and tf-LIFEs (Lago et al., 2007) were implanted longitudinally in the tibial nerve (group LIFE).

Stimulation was provided by a STIM'3D stimulator (Andreu et al., 2009), delivering series of monophasic rectangular pulses of 10 μs at 0.5 Hz, increasing the intensity in steps of 20 μA from 20 to 300 μA for TIME and LIFE and to 800 μA for cuff electrodes. Pulses were delivered through each of the active sites of the TIME and tf-LIFE against a small needle electrode placed near the nerve. For the cuff, stimulation was applied to each one of the four tripoles in the cuff, using the central pole as cathode and the outer poles as anodes.

The activation of tibialis anterior (TA), gastrocnemius medialis (GM), and plantar interosseus (PL) muscles was studied to verify if it was possible to stimulate one of the three muscles without producing significant excitation of the other two. The fibers innervating the TA muscle are located in the peroneal branch, whereas those innervating the GM and the PL muscles are in distinct locations in the tibial branch (Badia et al., 2010). The compound muscle action potentials (CMAPs) were recorded by small needle electrodes placed in each muscle, amplified and filtered (5 Hz, 2 kHz). The amplitude of the CMAP (M wave) was normalized to the maximum CMAP amplitude obtained for each muscle in the experiment. For each active site (as), k, a selectivity index (SIas) was calculated as the ratio between the normalized CMAP amplitude of that muscle, CMAPni, and the sum of the normalized CMAP amplitudes elicited in the three muscles (Veraart et al., 1993):

$$SIas_{i,k} = \frac{CMAPni}{\sum_j CMAPnj} \cdots \tag{6.1}$$

this index ranges from 0 (no activation of the target muscle) to 1 (activation of only the target muscle).

In order to compare the stimulation selectivity of the three devices, a selectivity index of the device (SId) was calculated. The SId was the product of the highest SIas for each muscle with one given electrode, and may range from 0.0307 (no selectivity; each SIas $=$ 0.333 for three muscles considered) to a maximal of 1.0 (maximal selectivity; each SIas $=$ 1).

$$SId = SI_{asPl} \times SI_{asGM} \times SI_{asTA} \cdots \tag{6.2}$$

6.2.1.2 Results

At low stimulation intensity selective stimulation of one muscle could be detected, due to the close contact of one electrode site with the corresponding muscular nerve fascicle. By progressively increasing the intensity of the

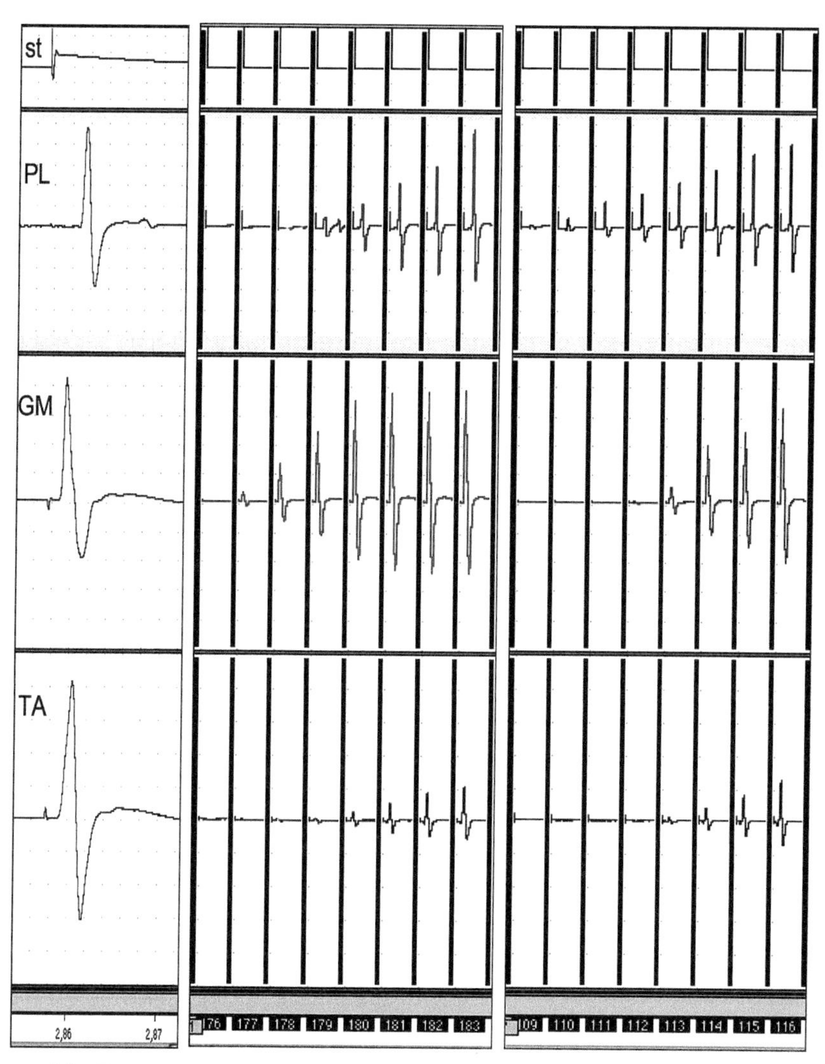

Figure 6.1 Examples of CMAPs recorded in plantar (PL), gastrocnemius medialis (GM), and tibialis anterior (TA) muscles with stimulation (st) at increasing pulse intensity, delivered from two different active sites of a TIME implanted in the rat sciatic nerve. Selective activation of GM (center block) and selective activation of PL muscle (right block) can be observed.

stimulus, the amplitude of CMAPs increased to reach the maximal value (i.e., activation of the whole muscle); however, the stimulus spread to other surrounding fascicles, therefore activating two or the three muscles tested (Figure 6.1).

The TIME device allowed selective activation of the PL, GM, and TA muscles when stimulating through different active sites. Selective activation using a multipolar cuff electrode was only possible for two muscles, GM and TA, whereas with the tf-LIFE it was not possible to activate selectively more than one muscle (either PL or GM for the implants in the tibial nerve). The intrafascicular electrodes TIME or LIFE provided excitation limited to the implanted fascicle, but did not extend stimulation to other nearby fascicles crossing the perineurial barrier (Badia et al., 2011).

The threshold for muscle activation was found at intensities between 24 and 66 μA similar for TIME and LIFE devices, and at higher levels ranging 180–330 μA with cuff electrodes. The mean thresholds with the TIME did not differ significantly between the three muscles. Cuff minimum threshold for the PL muscle was significantly higher than the threshold for GM and TA muscles, likely indicating a deep vs. superficial location of the corresponding motor fibers within the nerve.

With the LIFE implanted in the tibial fascicle good selectivity was obtained for the GM, moderately for the PL, but no activation of the TA muscle was possible. With the cuff electrode good selectivity was found for the GM and TA muscles, i.e., between different fascicles in the same nerve trunk, but it was not possible to separately activate PL from GM muscle, at the subfascicular level. In contrast, with the TIME different active sites selectively activated each one of the three muscles evaluated, indicating potential selectivity at fascicular and subfascicular levels (Figure 6.2). Calculation of the SId showed that the TIME has higher selectivity in the three muscles model tested than the cuff and LIFE electrodes.

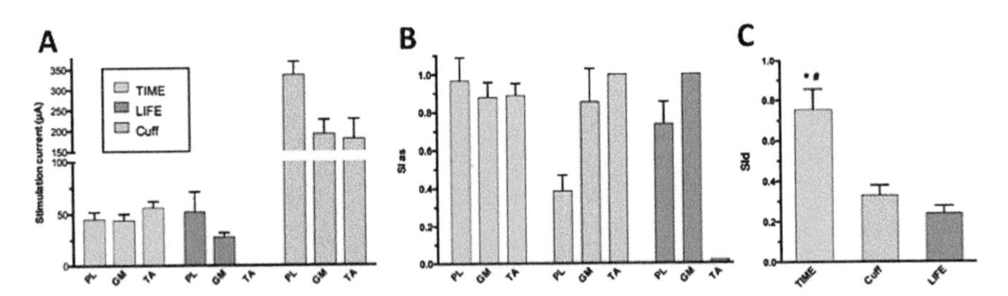

Figure 6.2 (A) Plot of the threshold of activation of the motor fascicles innervating PL, GM, and TA muscles, with TIME, LIFE, and cuff electrodes. (B) Plot of the Sias (best active site in each electrode) obtained for each of the muscles tested with TIME, LIFE, and cuff electrodes. (C) Plot of the SId corresponding to TIME, Cuff, and LIFE devices. Bars are mean and SEM. * $p < 0.05$ vs. Cuff; # $p < 0.01$ vs. LIFE. Data from Badia et al., 2011.

In order to evaluate the changes in selectivity depending on the relative location of the implanted TIME within the sciatic nerve, after performing the stimulation protocol with one TIME, we moved the TIME a distance of about 0.1 mm in the same direction that it was implanted and repeated the stimulation protocol. The comparison of the recruitment curves and the SIas showed variations as high as 100% in the SI for individual active sites between the two trials. This observation is a proof of concept that, once implanted, the intraneural TIME can be carefully repositioned to obtain the optimal stimulation selectivity of different fascicles and subfascicles in the nerve.

6.2.2 Recording Selectivity

To verify that the TIME was able also to serve as an adequate bidirectional interface, acute experiments were made on adult rats in which the sciatic nerve was implanted with a TIME for assessing its properties for recording neural activity. The aims of this work were: (1) to characterize the recording capabilities of the TIME in an experimental model, (2) to assess the spatial selectivity of the TIME for recording sensory neural activity at the sub-fascicular level, and (3) to assess the potential of TIME-recorded signals for discriminating neural activities evoked by stimuli of different sensory modalities (Badia et al., 2016).

6.2.2.1 Methods

The TIME was transversally inserted across the two main fascicles of the sciatic nerve, the tibial and peroneal branches, proximal to the knee, as reported in the previous section. The insertion was monitored under a dis-section microscope to ensure that the TIME active sites were located as: the first three of the right and left branches inside the tibial nerve, and the fourth ones within the peroneal nerve (Figure 6.3).

For assessing the spatial selectivity, compound nerve action potentials (CNAPs) were evoked from the digital nerves of toes 2 and 4, and recorded with the TIME. The digital nerves were stimulated by means of two small needles at the side of toes second and fourth, delivering pulses of 200 μs at 1 Hz, with increasing amplitude 1–10 V at steps of 1 V. Ten CNAPs obtained for each voltage pulse were averaged to determine the initial latency and the peak amplitude. The CNAP peak corresponding to Aαβ fibers was identified from its conduction velocity, and the mean maximal amplitude plotted as recruitment curves. A CNAP was considered when its amplitude exceeded two times the maximum value of the root mean square of the background noise. For each active site, a selectivity index (SIas) was calculated as the ratio between the CNAP amplitude exhibited for one stimulated digital nerve, CNAP toe i, and the sum of the CNAP amplitudes elicited in the two digital nerves (Veraart et al., 1993) as in the formula [1].

With the aim of assessing the capabilities of TIME for recording functional neural signals preferentially in different active sites and the possible discrimination of sensory modalities, two different experiments were performed. The first was designed to evaluate the spatial selectivity of afferent signals recorded with the TIME. For this purpose, a pressure stimulus was applied with a Von Frey filament on pad A, toe 2, and tip 2 in the medial side of the paw, and on pad D, toe 4 and tip 4 in the lateral side of the paw. The signal-to-noise ratio (SNR), expressed as the ratio of the signal RMS to the noise RMS, was calculated for each of the six active sites implanted within the tibial fascicle, that innervates the plantar surface of the rat hindpaw. In the second experiment, the objective was to assess the SNR values for stimuli of different sensory modalities. Three tactile, one proprioceptive, and one nociceptive mechanical stimuli were applied in this order: (1) soft tactile stimulus by brushing the sole with a small brush, (2) pressure on a plantar pad, (3) fast scratch on the midline of the hindpaw, (4) flexion of the toes of the paw, and (5) light prick of the plantar skin with a small needle (Raspopovic et al., 2010). Each test was repeated 10 times with intervals of at least 10 s between stimuli.

The recorded neural signals were then bandpass filtered by means of a finite impulse response (FIR) filter with 0.4 and 2.2 KHz cut-off frequencies, preserving the main power of the ENG signal. The mean absolute value (MAV) was extracted from the preprocessed data using a bin width of 100 ms and an overlap of 50% (Raspopovic et al., 2010). Finally, a Support Vector Machine (SVM) classifier was used to discriminate between the different stimuli. Data were divided in a Training set (Tr) consisting in a random selection of 50% of the trials, and in a Testing set (Tt) consisting in the rest of the data. The SVM classifier was trained with Tr using the optimal parameters, and tested with Tt during the discrimination of: (i) one stimulus vs. background activity, (ii) two stimuli and background activity, and (iii) three stimuli and background activity.

6.2.2.2 Results

Different active sites of the TIME selectively recorded the CNAP evoked by stimulating the digital nerves of toe 2 or toe 4, indicating that it was possible to obtain topographically selective recordings of neural signals from different targets at the subfascicular level with the TIME. The mean onset latency of the measured CNAP peaks was $\sim 2.0\,\mathrm{ms}$, which corresponds to a mean conduction velocity of $\sim 40\,\mathrm{m/s}$, as expected from a population of large myelinated mechanoreceptive fibers (Harper and Lawson, 1985). The stimulation threshold for recordings in AS1 and AS2 of both branches of the TIME was 5–7 V whereas for AS3 it was above 12 V. This fact is related with the position of the electrode inside the nerve trunk. Active sites placed near the target stimulated nerve bundle record the evoked CNAP at lower levels of stimulation than active sites placed at more distance that would require stimulation of more nerve fibers.

From the recruitment plots of the amplitude of the CNAP recorded in the tibial nerve with increasing magnitude of stimulation, we calculated the SIas. In three animals, it was possible to discriminate the CNAP coming from toe 2 and toe 4 with a SI higher than 71%. In other two rats, the SI was 100% for toe 2 but only 55% for toe 4, while in the remaining rat the SI was 73% for toe 2 and 56% for toe 4. The selectivity indices for active sites AS3, which were probably implanted in the region of the tibial nerve occupied by nerve fibers innervating muscles of the hindlimb, were lower than for AS2 and AS1, which were located within the subfascicles of the tibial nerve going to the plantar nerves (see Figure 6.3) (Badia et al., 2016).

During sensory stimulation of the plantar aspect of the hindpaw, we recorded bursts of action potentials from the TIME active sites. Afferent neural activity elicited by tactile, proprioceptive, and nociceptive stimuli did not show significant differences in the SNR level. The mean value of the SNR, being higher with the application of fast scratch stimulation than static pressure or pricking stimulation. For the same type of stimulus applied in one of the stimulation spots, the SNR of the evoked recordings varied in amplitude between different ASs; the SNR was highest at AS1, and lowest at AS3. These results are attributed to the location of AS1 closer to the bundle of sensory fibers that innervate the hindpaw than the other ASs (Figure 6.3). Thus, the TIME may give spatial discrimination of source of the recorded neural activity.

The capacity to discriminate different sensory activity was tested by using an automatized classifier. Considering the mean classification accuracy for the best channel in each of the rats, it was possible to robustly discriminate between different stimuli and background activity from the ENG signals recorded with the TIME. Mean values of 95.4%, 96.0%, and 85.2% were obtained when discriminating pressure, proprioceptive and nociceptive stimuli vs. background, respectively.

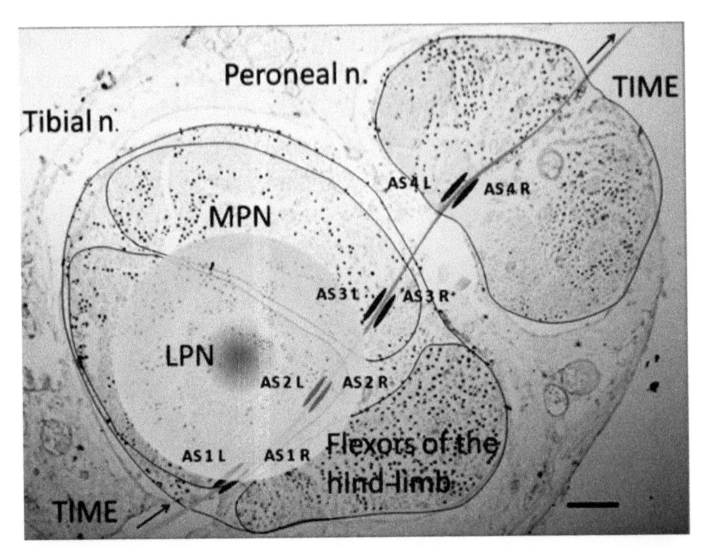

Figure 6.3 Microphotograph of a transverse section of a sciatic nerve immunolabeled against cholin-acetyl transferase (ChAT, dots stained in black) to label motor axons, and counterstained with hematoxilin to visualize the tissue. The narrow strip occupied by the intraneural portion of the TIME has been overlaid with a thin line (brown) and the active sites (AS, in black) marked in a possible position.

6.2.2.3 Discussion

The results of these works made on the rat model indicate that the TIME allows highly selective stimulation of different small muscular fascicles and even parts of the same fascicle within the sciatic nerve of the rat (Badia et al., 2011), and that it also offers good capabilities for selective recording of neural signals elicited by electrical and mechanical stimulation delivered at discrete areas of the foot of the rat (Badia et al., 2016). Furthermore, the post hoc processing of the signals recorded from the TIME allows the identification of the corresponding stimulus pertaining to different functional modalities. Relative comparison of the SI obtained suggests that the TIME may be more selective for stimulation of nerve fibers than for recording neural activity. The neural signal recorded is the superposition of all the afferent signals corrected by the distance factor, whereas in the case of stimulation, since an axon is not activated until the threshold of stimulation is reached, only the nearest axons to the active site will be activated at low levels of current, and therefore spatial selectivity will be present (Raspopovic et al., 2012).

Because of the careful insertion inside nerve fascicles, TIMEs and LIFEs achieved muscle activation at a considerably lower intensity than the extraneural cuff electrodes, further supporting the advantage of intraneural electrodes for reducing the amount of current needed for axonal stimulation. The intraneural placement of the active sites also allowed meaningful recordings of neural signals evoked by functional stimulation of afferent axons in the rat sciatic nerve. The recordings were usually of integrated action potentials, with seldom spikes, comparable to those obtained with other intrafascicular electrodes (Branner and Normann, 2000; Yoshida et al., 2000; Navarro et al., 2007).

In the comparative study for stimulation, the TIME was the only electrode tested that provided good stimulation selectivity at interfascicular and also at intrafascicular level with one electrode implanted, even in a small size nerve such as the rat sciatic nerve. In contrast, the LIFE allowed only selectivity at the intrafascicular level, and the multipolar cuff electrode only interfascicular selectivity. Thus, the TIME has advantage in the relation between selectivity and invasiveness. Even when the targeted muscle was stimulated above 30% of the maximal CMAP, a functionally relevant magnitude (Bao and Silverstein, 2005; Paternostro-Sluga et al., 2008), the SId of TIME remained higher than those of LIFE and cuff electrodes. These results suggest that the optimal application of TIME device may be in the portions of the peripheral nerve where different bundles of axons innervating different organs are in the inner part of the nerve forming or not a fascicle encircled by perineurium. By repositioning the TIME short distances in the transversal direction, the selectivity of single active sites may be considerably improved, suggesting the possibility of measuring the SIas of the TIME implant and then slightly adjust its position in the nerve to optimize stimulation outcome. Once the optimal position is achieved, the electrode can be fixed in place to the nerve to avoid undesired displacements.

The works here summarized provide evidence that the TIME is a good intraneural electrode to be used as a bidirectional interface between neuroprostheses and the peripheral nerves. The TIME may be appropriate for neuroprosthetic applications at mid and proximal levels of the nerves, reducing the number of electrodes to be surgically implanted and thus minimizing the risk of damage and the complexity of the interface connections.

6.3 Evaluation of TIME in the Pig Nerve Model

The pig nerve model was used for testing the TIME electrodes, because of pig nerves anatomical resemblance to humans nerves in terms of number of fascicles and nerve diameters. Two studies were conducted: (1) Initially acute studies were performed to develop and test implantation techniques and with the purpose to assess selectivity performance and required recruitment current of the TIME electrode; (2) Chronic studies were later conducted to assess the selectivity performance, recruitment current, stability and biocompatibility of the TIME (see chapter 5 also) during periods of implantation.

6.3.1 Acute Study of Stimulation Selectivity

TIMEs and tfLIFEs were implanted into the median nerve of the left foreleg of farm pigs (Kundu et al., 2014). The TIMEs were implanted transverse through the center of the nerve and the tfLIFEs were implanted parallel to the nerve direction. Patch electrodes were sutured to the seven muscles innervated by the median nerve distally for recording the evoked electromyogram (EMG). Stimulation was performed using monopolar rectangular 100 μs duration current pulses, gradually increased in intensity from 40 to 800 μA, with individual TIME/tfLIFE active sites as cathode. The evoked EMG responses from the monitored muscle were used as an indirect assessment of the nerve recruitment evoked by the tested TIME and tfLIFE electrodes. The EMG recruitment level in percentage (EMG_{RL}) for each muscle, was calculated by taking the normalized root mean square value of EMG response. The selectivity for each muscle (SI_m), was then calculated using the same selectivity index as in the rat experiments described above. As a constrain to reduce the influence of noise in the EMG recordings, SI_m was calculated only when the target muscle was activated $> 30\%$. The specific 30% limit was defined as a minimum requirement for the recruitment to be functionally relevant (Bao and Silverstein, 2005). To evaluate the performance of a whole TIME/tlLIFE electrode a device selectivity index (SI_d) was defined as the average of the maximal SI_m achieved for each of the seven monitored muscles. Finally, the required recruitment current was calculated as the average current used, when achieving the maximal SI_m.

After ending the stimulation protocol, the pig was euthanized. For eight TIMEs and four tfLIFEs the location of the electrode inside the nerve was determined via histology. For these instances the nerve was carefully dissected free and harvested after which it was frozen in liquid nitrogen, sectioned into 5 μm cross-sections and stained with Hematoxylin and Eosin.

6.3.1.1 Results

Results showed that individual muscles could be most selectively activated at the lower current levels. When using higher currents, the selectivity dropped, as more parts of the nerve were recruited. The TIME was on average capable of recruiting 2.2 ± 0.9 muscles to a $SI_m > 0.5$ whereas the tfLIFE on average only selectivly activated 1 ± 0.0 muscles. Overall the TIME had a higher SI_d

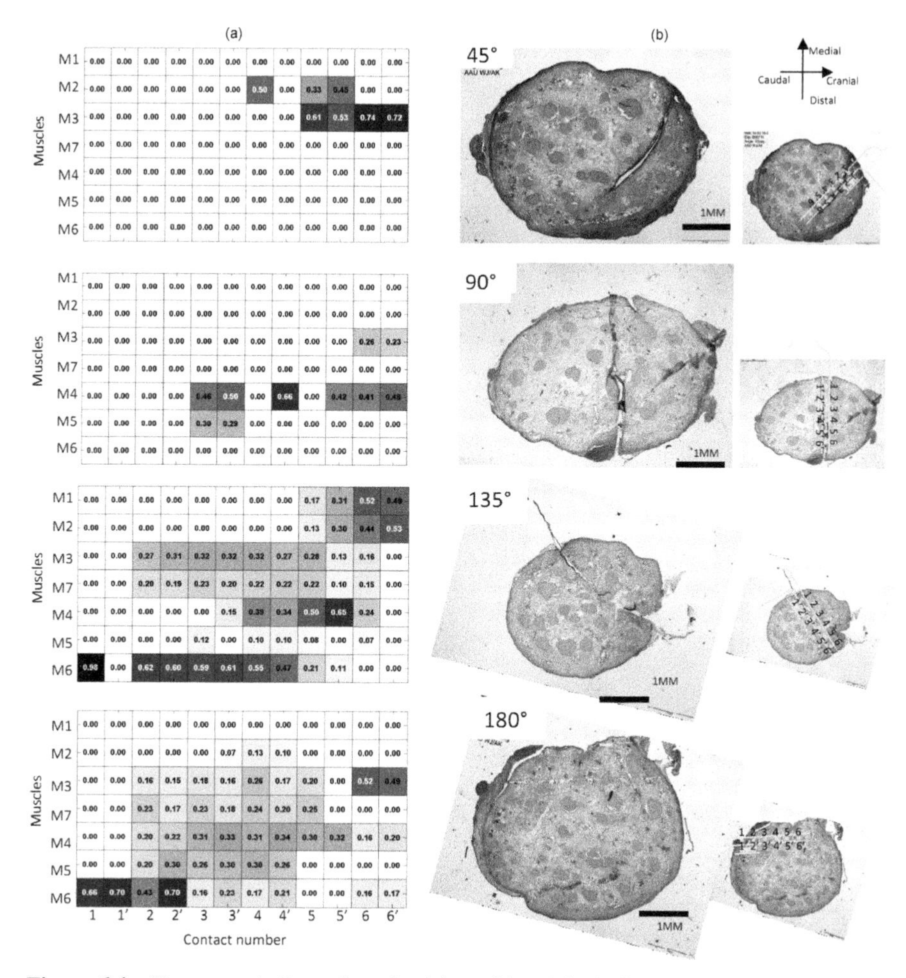

Figure 6.4 Heat maps indicate the selectivity achieved for individual muscles (M1–M7), 0 corresponding to white and black corresponding to 1, when using the different contact sites of the TIME (1–6 and 1'–6', corresponding to the contact sites on each side of the TIME loop structure) and tfLIFE (1,1–4,4'). The histology images show the corresponding traces of the inserted electrodes. In general, the TIMEs were better at activating several different muscles, whereas the tfLIFE tended to activate a single muscle selectively. Reprinted with permission from Kundu et al. 2014.

than the tfLIFE ($p = 0.02$) and performance was not found to be related to the angle with which the electrodes had been implanted into the nerve (see Figure 6.4). With respect to the required current, I_d, no differences were found between the two type of electrodes. Histological assessment, showed that all the electrodes had been placed between fascicles, and not inside the fascicles as intended.

6.3.2 Chronic Study of Stimulation Selectivity

Four female Göttingen mini-pigs (25–40 kg) were implanted with a total of six TIMEs over a period of >30 days (Harreby et al., 2015). TIMEs were always implanted through the center of the nerve; when implanting two electrodes in the same nerve, they were inserted at different angles to cover different subsections of the nerve. TIMEs were sutured to the epineurium using the anchoring points just at the entry point and at the ribbon, which connected it to the ceramic connector, see Figure 6.5A. The ceramic connector was fixed to tissue close to the nerve.

EMG patch electrodes were sutured onto five muscles innervated by the median nerve and a cuff electrode was attached to the median nerve to provide supramaximal stimulation.

The ceramic connector of the TIME implant was connected to a 16 channel circular connector (Omnetics Inc., Minneapolis, USA) via helically coiled MP35N wires enclosed in silicone tube shielding. The lead out wires from the EMG and cuff electrodes were also connected to a 16 channel circular connector. The silicone tube cables from TIMEs, cuff and EMG electrodes were tunneled subcutaneously to the back of the pig and mounted in a custom made stainless steel capsule, which was sutured to the skin.

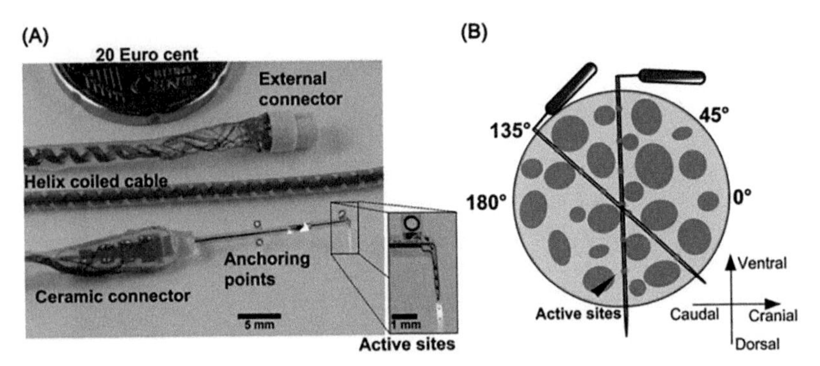

Figure 6.5 (A) picture of the TIME implant. (B) Illustration indicating the TIMEs were located inside the nerve. (B) Example was TIME electrodes are placed at 135° and 90°. As in the acute pig study, post-mortem findings showed the TIMEs had been located between the fascicles. Reprinted with permission from Harreby et al. 2014.

6.3.2.1 Follow-up methods

Follow-up sessions were performed two to three times a week for the duration of the implantation period. During these events stimulation and recording equipment was connected to the lead out wires via the Omnetics connectors at the back of the pig.

All stimulation was conducted at 2 Hz using rectangular 100 μs current pulses. Stimulation current gradually increasing up to 1200 μA, corresponding to the maximal charge capacity of the TIME (Boretius et al., 2012). A total of 22 stimulation configurations were evaluated for each TIME electrode: 12 corresponding to monopolar stimulation using each contact site (1–6 and 1'–6'). Then 10 nearest neighbor bipolar configurations were used (1–2 ...5–6 and 1'–2' ...5'–6').

In some instances EMG channels became unstable, in these cases monopolar EMG recordings were made when possible. If monopolar recording could not be satisfactorily obtained, the specific channel was omitted from the analysis.

During offline analysis three different techniques were applied to clean the EMG recordings for noise: (1) Stimulation artifacts were blanked out and replaced by a linear interpolation, (2) the signal was band pass filtered between 0.1 and 2 kHz, and (3) power line noise was reduced by subtracting a "quiet" but noisy recorded signal segment 20 ms (period of powerline noise) prior to the segment of interest. Evoked EMG activity was now quantified using the RMS value as in the acute rat and pig studies.

The required current for activating a particular muscle, ($I_{m30\%}$) was determined as the minimal current for any TIME contact configuration to obtain a recruitment level of 30% ($EMG_{RL30\%}$). The minimal required current for recruiting all monitored muscles using a single TIME device ($I_{d30\%}$), was simply calculated as the average $I_{m30\%}$ for each of the individual muscles monitored.

The selectivity index used in this study was based on the one used in the acute rat and pig study, but with a couple of adaptations/improvements. Initially the constant weighting SI ($\mathrm{SI_{CW}}$) was defined as:

$$\mathrm{SI}(I)_{\mathrm{CW,j}} = \frac{\mathrm{EMG}(I)_{\mathrm{RL,j}}}{\mathrm{EMG}(I)_{\mathrm{RL,j}} + (N_C - 1) * \left(\frac{\sum_{i=1|\mathrm{i} \neq j}^{N_A} \mathrm{EMG}(I)_{\mathrm{RL,i}}}{N_A - 1} \right)}$$

Here N_C, is a constant corresponding to the number of muscles intended to be monitored, and N_A is the number of muscles from which EMG could actually be recorded. This reduces the bias induced if EMG could not be recorded from all muscles, and thus the number of EMG channels changed. If $N_C = N_A$, $\mathrm{SI_{CW}}$ simply corresponds to the selectivity indices used in [1] and [2].

To ensure $\mathrm{SI_m} = 0$ when all muscles were equally activated, the following correction was applied to $\mathrm{SI_{CW}}$:

$$\mathrm{SI(I)_{m,j}} = \frac{N_C}{N_C - 1} * \left(\mathrm{SI(I)_{CW,j}} - \frac{1}{N_C}\right)$$

$$\text{if} \quad \mathrm{SI_{m,j}} < 0 \quad \text{then} \quad \mathrm{SI(I)_{M,j}} = 0$$

This function can be calculated for each contact combination. $\mathrm{SI_{m \geq 30\%,j}}$ was now defined as the highest $\mathrm{SI_m}$ value which could be achieved by a single TIME contact combination, when the recruitment level was $\mathrm{EMG_{RL30\%}}$ or above. A muscle was defined as being selective activated when $\mathrm{SI_m} > 0.4$, corresponding to an average nontarget recruitment of 1/4 of that of the target muscle. To quantify the overall selectivity performance of a whole TIME device the $\mathrm{SI_{d>30\%}}$ was defined as the mean of all $\mathrm{SI_{m \geq 30\%}}$ values.

6.3.3 Results

Six TIMEs were implanted in four pigs for a total of 33.8 ± 2.4 days. Current pulses evoked EMG responses, occurring mainly between 3 and 10 ms after the stimulation pulse onset, see Figure 6.6. Recruitment of muscles in general started at the same current level, but individual muscles often differed in how they were recruited at higher currents, these differences meant that some muscles became more selectively recruited than others. Recruitment curves were in general not smooth sigmoid curves, but rather increased with a mix of increases and plateaus (see Figure 6.7b, c). In general monopolar stimulation evoked more muscle activity than bipolar stimulation did at similar current levels. During the first follow-up around half of the bipolar configurations were able to evoke $\mathrm{EMG_{RL}} > 30\%$, however, during the final follow-up session only 3/60 could evoke such EMG. Initially, in five out of six TIMEs all channels could evoke at least one EMG channel to $\mathrm{EMG_{RL}} > 30\%$. Two TIME electrodes gradually stopped being able to recruit muscles, however, the remaining electrodes could recruit to $\mathrm{EMG_{RL}} > 30\%$ on nearly all contact sites until last follow-up. The average $I_{d30\%}$ was initially 488 ± 68 μA and increased to 769 ± 128 μA during the last follow-up.

The $\mathrm{SI_{d \geq 30\%}}$ started at 0.25 ± 0.04 during the initial follow-up and gradually decreased to 0.14 ± 0.05 during the last follow-up ($P = 0.12$).

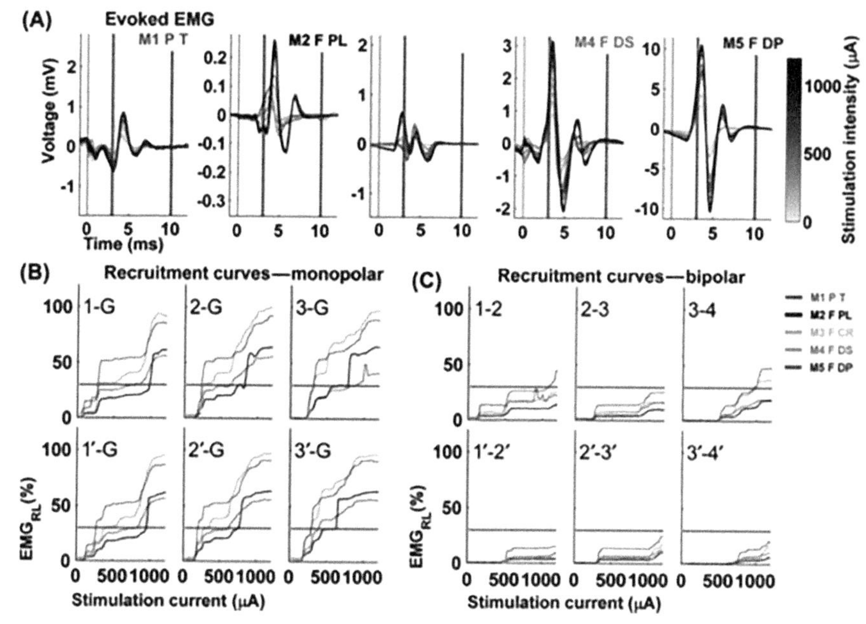

Figure 6.6 (a) The raw evoked EMG response when stimulating in P2T1 at day 7 from the five monitored muscles when stimulating monopolar using 3'. The dotted vertical lines indicate 30% EMG recruitment ($\mathrm{EMG_{RL30\%}}$). (b) Shows the recruitment $\mathrm{EMG_{RL}}$ for each of muscles during monopolar stimulation (G–Ground) with a subset of six contact sites of P2T1. Note that the recruitment curves are not smooth, but rather have consist of steep increases and plateaus. (c) Shows the recruitment curves related to a subset of bipolar stimulation configurations. Note that the recruitment level is significant lower for bipolar stimulation than for monopolar stimulation. Reprinted with permission from Harreby et al. 2014.

Based on our definition of selectively activated muscles, each TIME could initially activate 1.17 ± 0.37 muscles, which dropped to 0.67 ± 0.38 during the last follow-up ($P = 0.18$). Neighboring contact sites on the TIME (same side and on opposite sides), tended to recruit the same muscles, with only small variations in the recruitment curves. Electrodes P2T1 and P2T2 which were both implanted in the same nerve, recruited different subsets of muscles (Figure 6.7a, b). During the initial follow-up bipolar stimulation configurations were selected at the best stimulation configuration in 1/3 of cases, when calculating the $\mathrm{SI_{d\geq30\%}}$, however, if these configurations were left out of the overall $\mathrm{SI_{d\geq30\%}}$ calculation (i.e., based only on monopolar stimulation) the value dropped only by 0.005, indicating that bipolar configurations contributed insignificantly to the selectivity.

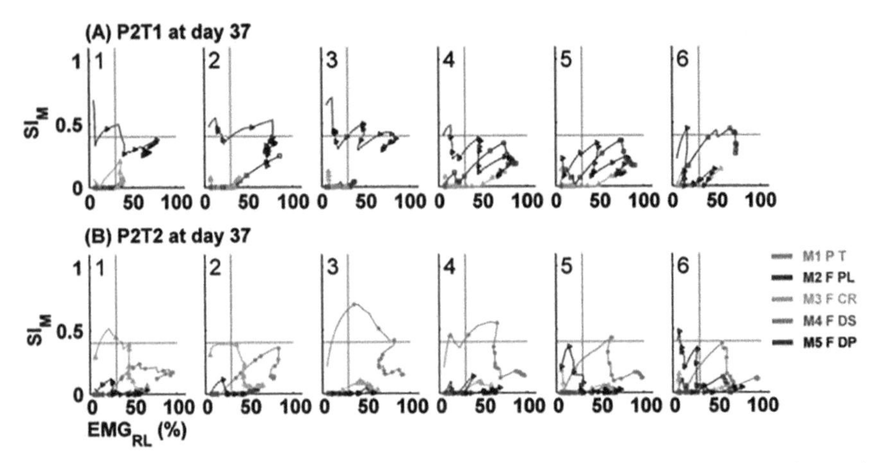

Figure 6.7 The selectivity of individual muscles is shown as a function of muscle recruitment level for a subset of monopolar stimulation configurations from electrodes P2T1 and P2T2 during the last follow-up session at day 37. The vertical and horizontal dotted lines indicate the limits for $\mathrm{EMG}_{\mathrm{RL}30\%}$ and $\mathrm{SI_m} = 0.4$, thus based on our definitions a muscle is selectively recruited if it enters the upper right quadrant. In P2T1 muscles: M5 and M2 are selectively activated, in P2T2 M3 and M1 are selectively recruited. Reprinted with permission from Harreby et al. 2014.

6.3.4 Discussion

The acute pig experiments showed that the TIME design was more effective than the tfLIFE to selectively activate different parts of a large polyfascicular nerve. This is not surprising as the tfLIFE was designed to be placed in parallel to the nerve, meaning that all the contact sites of one tfLIFE would be placed close to the same few fascicles in the nerve. In contrast, the TIME was implanted transversal through the nerve, which means that the contact sites were placed along the whole cross-section of the nerve.

Although different methodologies were used and thus it is difficult to compare directly, the selectivity seen in the pig model seem to be lower as those achieved by implanting TIMEs in rats (Badia et al., 2011). There may be several reasons for this. Badia et al. were able to insert the TIMEs directly into the target fascicles of the small rat nerve, this resulted in a much smaller recruitment current and a better selectivity. In the large nerve of the pig it was not possible for us to insert into specific fascicles. Furthermore, the location of the specific fascicles which innervated the muscles from which EMG were recorded was not known. This meant that electrodes were inserted "blindly" through the epineurium. Inside the nerve, the needle then glided of the tough perineurium of individual fascicles which resulted in the extrafascicular placement. This location both increased the required recruitment current needed and reduced the achievable selectivity. Methods to overcome the extrafascicular location in the future could be to make previous neurolysis of the nerve to separate the fascicles or to insert the guide needle at high velocity, as have been done when placing the Utah array (Warwick et al., 2003).

References

Andreu, D., Guiraud, D. and Souquet, G. A. (2009). Distributed architecture for activating the peripheral nervous system. J. Neural Eng. 6:026001.

Badia, J., Boretius, T., Andreu, D., Azevedo-Coste, C., Stieglitz, T. and Navarro X. (2011b). Comparative analysis of transverse intrafascicular multichannel, longitudinal intrafascicular and multipolar cuff electrodes for the selective stimulation of nerve fascicles. J. Neural Eng. 8:036023.

Badia, J., Pascual-Font, A., Vivó, M., Udina, E. and Navarro, X. (2010). Topographical distribution of motor fascicles in the sciatic and tibial nerve of the rat. Muscle Nerve. 42:192–201.

Badia, J., Raspopovic, S., Carpaneto, J., Micera, S. and Navarro, X. (2016). Spatial and functional selectivity of peripheral nerve signal recording with the transversal intrafascicular multichannel electrode (TIME). IEEE. Trans. Neural Syst. Rehabil. Eng. 24:20–27.

Bao, S. and Silverstein, B. (2005). Estimation of hand force in ergonomic job evaluations. Ergonomics. 48:288–301.

Boretius, T., Badia, J., Pascual-Font, A., Schuettler, M., Navarro, X., Yoshida, K. and Stieglitz, T. (2010). A transverse intrafascicular multichannel electrode (TIME) to interface with the peripheral nerve. Biosens Bioelectron. 26:62–69.

Boretius, T., Yoshida, K., Badia, J., Harreby, K. R., Kundu, A., Navarro, X., Jensen, W. and Stieglitz, T. (2012). A transverse intrafascicular multichannel electrode (TIME) to treat phantom limb pain – Towards human clinical trials, *Proceedings of 2012 4th IEEE RAS & EMBS International Conference on Biomedical Robotics and Biomechatronics*, pp. 282–287.

Branner, A. and Normann, R. A. (2000). A multielectrode array for intrafascicular recording and stimulation in sciatic nerve of cats. Brain Res Bull. 51:293–306.

Delgado-Martinez, I., Badia, J., Pascual-Font, A., Rodriguez-Baeza, A. and Navarro, X. (2016). Fascicular topography of the human median nerve for neuroprosthetic surgery. Front Neurosci. 10:286.

Dhillon, G. S., Kruger, T. B., Sandhu, J. S. and Horch, K. W. (2005). Effects of short-term training on sensory and motor function in severed nerves of long-term human amputees. J. Neurophysiol. 93:2625–2633.

Gustafson, K. J., Pinault, G. C., Neville, J. J., Syed, I., Davis, JAJr., Jean-Claude, J. and Triolo, R. J. (2009). Fascicular anatomy of human femoral nerve: Implications for neural prostheses using nerve cuff electrodes. J. Rehabil. Res. Dev. 46:973–984.

Harper, A. A. and Lawson, S. N. (1985). Conduction velocity is related to morphological cell type in rat dorsal root ganglion neurones. J. Physiol. 359:31–46.

Harreby, K. R., Kundu, A., Yoshida, K., Boretius, T., Stieglitz, T. and Jensen, W. (2014). Subchronic stimulation performance of transverse intrafascicular multichannel electrodes in the median nerve of the Göttingen minipig. Artif Organs; 39:E36–E48.

Kundu, A., Harreby, K. R., Yoshida, K., Boretius, T., Stieglitz, T. and Jensen, W. (2014). Stimulation selectivity of the thin-film longitudinal intrafascicular electrode (tfLIFE) and the transverse intrafascicular multi-channel electrode (TIME) in the large nerve animal model. IEEE Trans. Neural Systems Rehab. Eng. 22:400–410.

Lago, N., Yoshida, K., Koch, K. P. and Navarro, X. (2007). Assessment of biocompatibility of chronically implanted polyimide and platinum intrafascicular electrodes. IEEE Trans. Biomed. Eng. 54:281–290.

Navarro, X., Lago, N., Vivó, M., Yoshida, K., Koch, K. P., Poppendieck, W. and Micera, S. (2007). Neurobiological evaluation of thin-film longitudinal intrafascicular electrodes as a peripheral nerve interface. Proc IEEE 10th International Conference on Rehabilitation Robotics, pp. 643–649.

Navarro, X., Valderrama, E., Stieglitz, T. and Schuettler, M. (2001). Selective fascicular stimulation of the rat sciatic nerve with multipolar polyimide cuff electrodes. Restor Neurol Neurosci. 18:9–21.

Paternostro-Sluga, T., Stieger, M. G., Posch, M., Schuhfried, O., Vacariu, G., Mittermaier, C., Bittner, C. and Fialka-Moser, V. (2008). Reliability and validity of the Medical Research Council (MRC) scale and a modified scale for testing muscle strength in patients with radial palsy J. Rehabil. Med. 40:665–671.

Raspopovic, S., Carpaneto, J., Udina, E., Navarro, X. and Micera, S. (2010). On the identification of sensory information from mixed nerves by using single channel cuff electrodes. J. NeuroEng. Rehabil. 7:17.

Raspopovic, S., Capogrosso, M., Badia, J., Navarro, X. and Micera, S. (2012). Experimental validation of a hybrid computational model for selective stimulation using transverse intrafascicular multichannel electrodes. IEEE Trans. Neural Systems Rehabil. Eng. 20:395–404.

Yoshida, K. and Jovanovic Stein, R. B. (2000). Intrafascicular electrodes for stimulation and recording from mudpuppy spinal roots. J. Neurosci. Methods. 96:47–55.

Veraart, C., Grill, W. M. and Mortimer, J. T. (1993). Selective control of muscle activation with a multipolar nerve cuff electrode IEEE Trans. Biomed. Eng. 40:640–653.

Warwick, K., Gasson, M., Hutt, B., Goodhew, I., Kyberd, P., Andrews, B., Teddy, P. and Shad, A. (2003). The application of implant technology for cybernetic systems. Arch. Neurol. 60:1369–1373.

7

Synchronous Multichannel Stimulator with Embedded Safety Procedure to Perform 12-Poles TIME-3H 3D Stimulation

David Andreu[1], Pawel Maciejasz[1], Robin Passama[1], Guy Cathebras[1], Guillaume Souquet[2], Loic Wauters[2], Jean-Louis Divoux[2,*] and David Guiraud[1,*]

[1]LIRMM, University Montpellier, INRIA, CNRS, Montpellier, France
[2]Axonic, MXM, France
E-mail: david.guiraud@lirmm.fr
*Corresponding Authors

7.1 Introduction

The human functions are controlled through different ways among which both central and peripheral nervous systems (CNS, PNS) play a major part. Within this complex network, the information code is based on action potential (AP) generation. Then, when a sensory function is deficient, inducing APs provides for a partial restoration of the lost function.

Electrical stimulation (ES) induces APs by depolarization of the membrane of the targeted cell, i.e., axons or muscle fibers. From the fifties, ES has been successfully used in a growing set of applications linked to motor and sensory impairments including pain management.

Attempts to use this approach have been made in movement rehabilitation, such as drop foot syndrome for hemiplegic patients (Liberson et al., 1961) and more complex movements or functions for para- and quadriplegic patients (Kralj et al., 1989; Davis et al., 1997; Kobetic et al., 1997, 1999; Rijkhoff, 2004; Guiraud et al., 2014). In the latter case, the functional results could prove to be substantial, including, for instance, recovery of the grasp function for quadriplegic patients, who might then be able to grab hold of objects, eat, and even, in the best cases, write with a pen (Smith et al., 1998).

In the sensory area, cochlear implants allow to recover sound perception for deep deaf persons. The principle is basically the same as for motor restoration with a set of electrodes that increased from one to more than twenty over the thirty past years, located in the cochlea in order to activate the remaining auditory neural circuits (Djourno et al., 1957). It allowed equipped deaf people to understand speech and in the best case speak as healthy subjects.

However, available implanted stimulators dedicated to humans, remained too limited to explore widely all the possibilities that new techniques could provide (Mastinu et al., 2017). In particular, the selective approach applied to nerves through either multicontact cuff (Schiefer et al., 2008) (Dali et al., 2018) or, in the present work, intra fascicular electrodes, needs for advanced multichannel stimulators. The TIME approach, proved to be quite efficient (Badia et al., 2011) but needs for a synchronous 12 outputs stimulator design. The selectivity was a mandatory feature as the project aims at inducing very specific sensation's feedback whereas the whole nerve obviously contains multiple sensory targets, but also, even unused, motor ones. Moreover, as the stimulator was used on humans, intrinsic safety was further needed. The chapter describes the design, the use and the main important results issued from this first-in-man test with multicontact TIME electrode.

7.2 Bench-top Stimulator

7.2.1 Design of the Bench-top Stimulator (Stim'ND)

Stim'ND is based on two main parts, a digital part and an analog one. The analog part of Stim'ND is dedicated to stimulus generation and the digital part, corresponding to the stimulation controller, embeds the functionalities required for the stimulation execution and its remote programming and control. This Stim'ND, whose global architecture is schematically represented in Figure 7.1, relies on the FES distributed architecture paradigm consisting in decentralizing stimulation control close to the electrodes (Andreu et al., 2009).

7.2.1.1 From specifications to design of the stimulator

Globally, the stimulator must allow the generation of multichannel current pulse stimulus patterns, with the possibility to modify waveform, amplitude, pulse duration, pulse rate, number of pulses, number, and timing of active

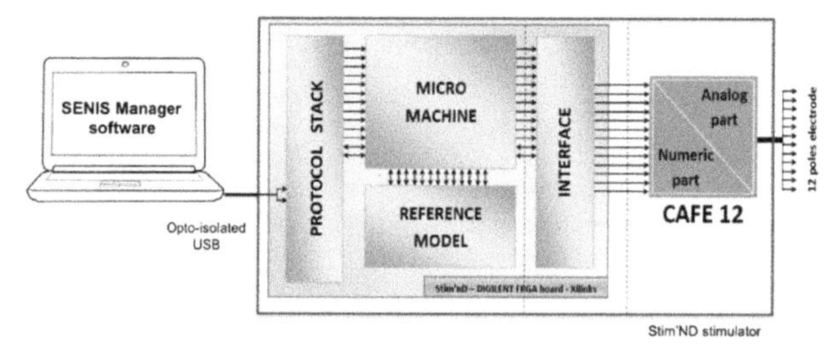

Figure 7.1 Schematic representation of Stim'ND architecture.

Table 7.1 Stimulator specifications

Stimulation Parameter	Range	Step Size
Pulse waveform	Fully programmable	1 μs step
Pulse amplitude	5 mA	1.3 μA (using 1/15 ratio) 20 μA without ratio programming
Pulse duration	Fully programmable	1 μs
Time between two stimulation sequences	Fully programmable	1 μs
# Pulses in a pulse train	Fully programmable	Application software dependent
Frequency of pulses in a pulse train	Fully programmable	Single pulse mode, 10–500 Hz
# Sites activated in full synchrony	Up to 12	Individually as anode and cathode but at least one of each type, total simultaneous current limited to 5 mA

channels. Moreover, the modification of the pattern's amplitude and duration, must be allowed "on-line," i.e., while the stimulation is running. This functionality is called real-time modulation and is performed using an experimenter-dedicated computer.

From a technical point of view, its specifications are given in Table 7.1. Since the stimulator must be able to execute different kinds of stimulation waveforms on different electrode configurations, we thus designed it according to two main concepts: 3D electrodes and microprograms (MPs).

3D electrode: The stimulator drives the thin-film intrafascicular multichannel electrode array (TIME-3H) with a maximum number of simultaneous poles to be driven set to 12. Since active channels can correspond to any TIME-3H configuration, the stimulus generator must be fully configurable, both in terms of poles' electrical configuration (active poles and their polarity) and in terms of accurate current repartition between active poles. Such a configuration is called "3D electrode" (VE). A VE is thus an entity that can be specified through the software environment and executed, from an analog device configuration point of view, by the stimulator. The possibility to specify different VEs is of importance since for an electrode placed within the same nerve a number of possible combinations of active sites are supposed to induce different sensations. This feature is based on the hypothesis that a somatotopy exists within the targeted peripheral nerve.

Microprogram: Since several waveforms can be tested, from a simple pulse to a pulse train and even other waveforms than pulses, the stimulator must offer a simple way to program pulses, ramps, and any combination of them to define more complex waveforms like trapezoidal waveforms or approximated exponential ones. Moreover, the polarity of the waveform is easily defined to provide charge-balanced waveforms. The stimulator being remotely programmable, a stimulus (i.e., a sequence of waveforms) is described by means of a MP.

Thus the stimulator, via its embedded specific processor (cf. Section 7.2.1.3), must offer a set of instructions that eases the description of a stimulation profile. Since some parameters of the waveform must be modulated in real-time during experiments, to find sensation and pain thresholds, parameters of the stimulus – duration, amplitude, time between pulses, and time between two stimulus (stimulation frequency) – must be externally modifiable parameters. That means that these parameters will have an initial value specified by the corresponding instruction of the MP and an effective value modified by remote control, at any time while stimulation is running.

Since some parameters of the waveform can be modulated in "real-time" during experiments, the security must be ensured on both sides:

- First, a security module embedded in the stimulator must ensure that a specified set of constraints will always be respected. If not, this module must immediately stop the stimulation and notify the detected violation to the operator. Specified constraints will concern, for instance, the maximum charge quantities that can be applied; these constraints will be defined prior to the use of the stimulator, i.e., parameters cannot be modified once the stimulation is running.
- Second, from the control environment in order to limit the range of amplitude and duration values that can be applied; this feature aims at ensuring comfortable and painless sensations.

Since different stimulus (MP) and multisite stimulations (VE) can be achieved for sensation comparison purposes, the stimulator should allow quick swapping between two multisite stimulations to ease experiments.

So, the stimulator offers the possibility to remotely execute and switch between two MP/VE (one at a time). Finally, since the stimulator must be remotely controlled, it embeds a communication module.

7.2.1.2 The Stimulus Generator

Concerning the analog front-end, the basic idea consists in providing synchronous current sources for each pole. The new idea we introduced (Guiraud, 2011) is based on the decoupling of the ratio and the global current control. To do so, a unique digital to analog current converter (DACC) is used followed by a programmable current divider. In this case the current is controlled in a completely different manner.

The most interesting benefits of such structure are that it allows:

- A constant step size of the global current amplitude *Iref* and a constant maximum amplitude *255*Iref*, whatever the ratios are if their sum is constrained to 16.
- An independent control of ratios and global amplitude.

However, equivalent structures between both solutions may be found but the cost is high. In a solution with 12 DACC, if we want to guarantee the specifications in the worst case, we have to provide for DACC with 12-bit resolution that leads to 144 bits word length. Moreover, DACC are much more power supply demanding than dividers. An ASIC for a 12-pole structure has been developed (Figure 7.1). This structure is able to provide sinking or sourcing currents in order to allow each pole to be configured either as a cathode or an anode (Figure 7.2). The programmed current divider at each output stages, allows to map the VE concept to the hardware directly.

Indeed, this advanced output stage allows for a higher level of software abstraction. In fact, on a functional point of view, both features mean different things. The ratios configuration determines on which zone of the nerve the current lines that induce action potentials are focused, whereas the global amplitude is linked to the extent of this area around the focus point. Our structure further optimized software design and data-flow.

7.2.1.3 The stimulation controller

The stimulation controller is the digital part of the stimulator. Its architecture relies on a set of interconnected modules; each one corresponding to previously mentioned functionalities.

The **execution module**, called "micromachine," can be seen as a specific and very small processor, similar to an application-specific instruction-set processor (ASIP) that runs MPs written in a FES-dedicated, reduced 32-bit

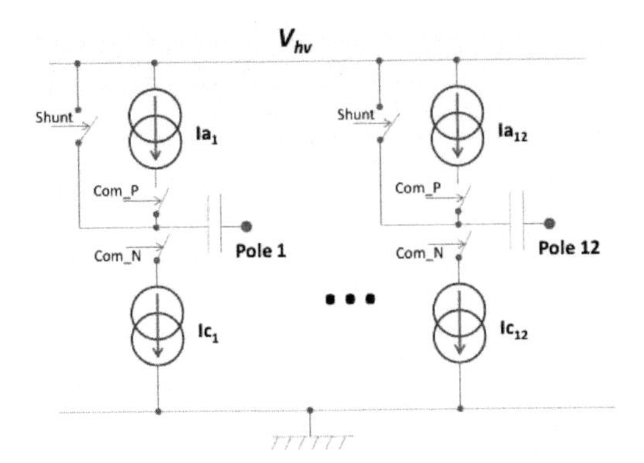

Figure 7.2 Principle of the output stage. Each channel can be configured as shunt (anode) for passive discharge, anode controlled current or cathode controlled current. One current source is used and spread over the 12 poles through ratios (Ia_i, Ic_i).

instruction set (summarized in Table 7.2). Doing so, it set the output configuration (VE) and drives the analog subsystem by calibrating the current pulse (waveform, amplitude, duration) to be applied to the multipolar electrode (Figure 7.3).

The instruction set is limited to four basic instructions:

- *SIT*. This instruction generates a positive or a negative pulse with a given intensity *I* during a given time *T*. A *RT_SIT* instruction is a *SIT* where *T* and *I* can be remotely modulated (during execution). In Figure 7.4, the seventh instruction is a *SIT* generating a positive pulse of 5 mA amplitude and 300 μs width.
- *RP_SIT*. This instruction generates a positive or negative ramp with a given step magnitude *I*, a given step duration *T*, during a number of steps *N*. It can also indicate that the starting intensity is dependent on the preceding instruction. A *RT_RP_SIT* instruction is a *RP_SIT* instruction where *T* and *I* can be remotely modulated. In Figure 7.4, the third instruction is a *RP_SIT* generating a train of 21 positive pulses (steps) with a 200 μA variation of amplitude and a 10 μs pulse width.
- *NST*. This instruction sets the high impedance state or generates a limited passive discharge on active anodes and cathodes during a given time *T*. A *RT_NST* instruction is a *NST* where *T* can be remotely modulated. In Figure 7.4, the ninth instruction is a *NST* enabling a passive discharge during 20 ms.
- *Loop*. A *Loop* instruction is interpreted as a termination marker for the MP. At termination it generates a limited passive discharge followed by a

Table 7.2 Instruction set of the Stim'ND micromachine

Name	Description	Possibilities	Parameters
NST	Discharge instruction	Passive discharge or Inter stimulation	Time (0–16 777 215 μs)
SIT	Basic stimulation instruction		Intensity (0–255 μA) Time (0–511 μs)
LOOP	Loop instruction	Finite or infinite Passive discharge or Inter stimulation	1–32 768 loop
Conf_P	Pole configuration	Anode, cathode, shunt (anode), high-impedance	
Conf_R	Ratio configuration	Ratios $K/16$	K (1–16)
RT_NST	Real-time discharge instruction	Passive discharge or interstimulation	Time (0–16 777 215 μs)
RT_SIT	Real-time stimulation instruction	Modulation of I and T	Intensity (0–255 μA) Time (0–511 μs)
RP_SIT	Ramp stimulation instruction	Current intensity dependence with previous instruction	Initial intensity (0–255 μA) Initial time (0–255 μs) Step number (0–255) Intensity step (0–255 μA) Time step (0–255 μs)
RT_RP_SIT	Real-time ramp stimulation instruction	Current intensity dependence with previous instruction Modulation of I and T	Initial intensity (0–255 μA) First step duration (0–255 μs) Number of steps (0–255) Intensity step (0–255 μA) Time step (0–255 μs)

global passive discharge for all poles or puts the poles in high impedance state. It also embeds a jump operation addressing a previous instruction in the sequence, to generate a loop in the MP. This back jump can be repeated many times or infinitely until program explicit stop, as it is the case of the last instruction of the MP given Figure 7.4.

Two other instructions are used to configure the analog output stage, since this configuration is separated from the stimulation profile itself:

- *CONF_P* is the instruction used to configure active poles. For each pole to be used, it defines whether a pole is a cathode, an anode, a shunt (not controllable anode) or is in high impedance state. The unique *CONF_P* in Figure 7.4 specifies that poles 1 and 3 are cathodes and pole 2 is an anode.

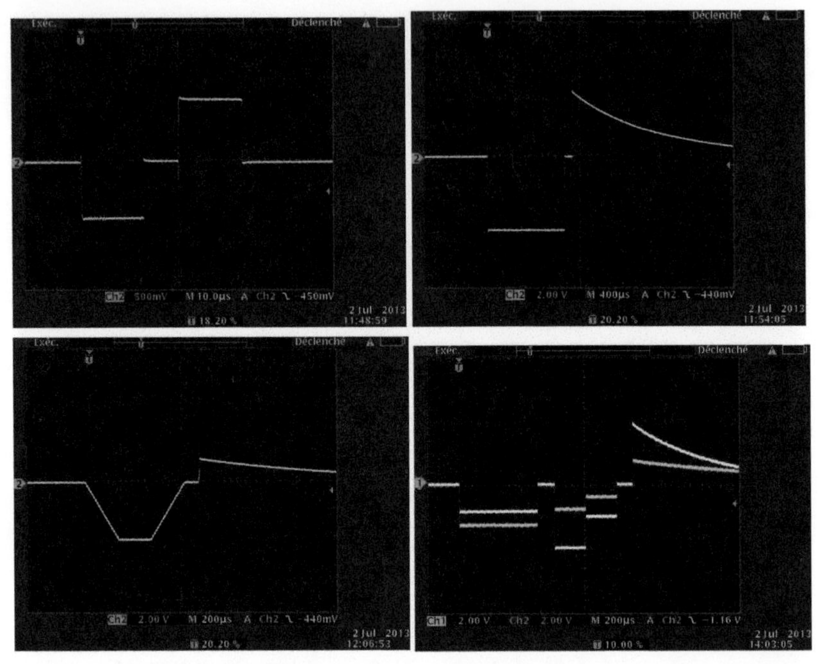

Figure 7.3 Four different stimulation waveforms generated by the miniaturized stimulator in bipolar mode (left-up). Rectangular biphasic charge balanced waveform (20 μs, 1 mA) with interstim (right-up) biphasic with passive discharge (1 ms, 4 mA) (left-down) biphasic trapezoïdal pulse with passive discharge. Train of pulses on a tripolar configuration with different current ratios, followed by a passive discharge (right-down). The signal is generated on a 1 kΩ resistor.

Inst		I	T	Steps	Note
CONF P	Cathodes : 1, 3, anode : 2				
CONF R	K1 : 75 %, K3 : 25 %, A2 : 100 %				
RP_SIT	+	200 μA	10 μs	21	
NST			20 μs		interstim
RP_SIT	-	200 μA	10 μs	21	
NST			100 μs		discharge
SIT	+	5 mA	300 μs		
NST			20 μs		interstim
SIT	-	5 mA	50 μs		
NST			100 μs		interstim
NST			20 ms		discharge
LOOP					endless

Tek Exéc. Déclench
Ch1 2.00 V Ch2 10.0 V M 200µs A Ch2 ƒ 2.40 V
Ch3 2.00 V

Figure 7.4 Example of 48-byte MP (left). Resulting stimulus with ch1 being cathode 1, ch3 cathode 3, and ch2 a trigger (right).

- *CONF_R* is the instruction used to configure ratios of current between active poles (controlled anodes and cathodes). In Figure 7.4, the unique *CONF_R* specifies that 75% of the current will be generated by cathode 1 and 25% by cathode 3.
- If *CONF_P* and *CONF_R* instructions must be used at the beginning of MPs, they may also be inserted anywhere else in the sequence of instructions to dynamically reconfigure poles and ratios, in order to apply different waveforms to the selected poles.

The **interface module** supports the interface with the stimulus generator. It ensures the effective electrical configuration of the analog device corresponding to the VE on which the stimulation has to be applied. It also maps control signals used by the micromachine with input signals of the analog device.

The **monitoring module** is based on a reference model in charge of monitoring in respect of neurophysiological constraints. These constraints are described Table 7.3; for instance, *Qmax* is set to 120 nC which is the safe limit of charge injection that the TIME-3H electrode can support. The monitoring of these constraints is based on simultaneous application of the stimulation to the nerve (via the stimulus generator) and to the reference model. Each time the micromachine sends a command to the stimulus generator, it sends the same command to the reference model. If any constraint is violated, the reference model forces the micromachine to immediately stop the stimulation and commute to a safe state. Thus, any error detection puts the stimulator into a safe mode regarding the physiological system under control (putting the output stage into a discharge mode). The micromachine then must be remotely rearmed by the experimenter in order for the stimulation to be once again authorized.

The stimulator does not perform any embedded diagnostic; it only indicates the type of error that has been detected. Note that any error due to the MP itself (e.g., coding error) is directly detected and managed within the execution module.

The **communication module** contains a three-layer protocol stack structured according to the reduced OSI-model with application, MAC (medium access control) and physical layers. The MAC layer provides logical addressing (and thus packet filtering) and deterministic medium access control mechanisms, based on a master/slave model (Godary et al., 2013). The application layer decodes the incoming packets, executes

Table 7.3 Parameters of the monitoring module

Name	Description	Parameters
QMax	Maximum quantity of electrical charges injected by "stimulation" (SIT > 0), or by "active discharge" (SIT < 0)	Electrical charges limit (0–6.5535 μC) with a 100 pC step
QThreshold	Maximum accepted residual quantity of electrical charges after passive discharge (after the end of a microprogram)	Electrical charges limit (0–6.5535 μC) with a 100 pC step
Min LPD duration	Minimum passive discharge time (after the end of a microprogram)	Duration limit (0–65 535 μs)

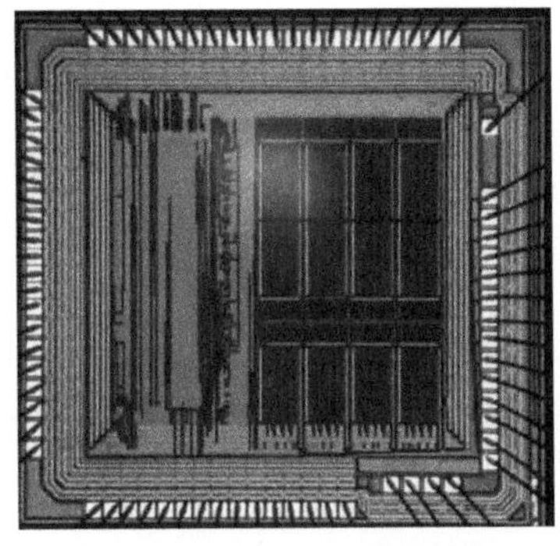

Figure 7.5 12-pole ASIC that can be seen in the center, about 4 mm × 4 mm, 0.35 μ HV technology.

the request by controlling the micromachine and the reference model, and it also encodes responses. From an application point of view, the application layer supports all the communication with the experimenter-dedicated computer (cf. Section 7.3.1): download/upload of MP and VE, start/stop/commute/rearm of MP, notification of error detection, monitoring parameters' initialization, and remote modulation of stimulation parameters.

7.2.2 Prototyping of the Stimulator

The prototype of bench-top stimulator Stim'ND is essentially based on the assembly of two chips, a digital chip and an analog one. The analog chip is an ASIC (Figure 7.5) and the digital one is an FPGA, both being detailed afterward.

7.2.2.1 Prototyping of the stimulus generator

This is the heart of the device. It is connected to the patient via the charge balancing capacitors and the electrodes. It consists in analog stages only, and mainly a digitally driven current sources and digitally configurable switches. The current sources are based on a patented method (Guiraud et al., 2010) that allows performing synchronous multipolar stimulation with fixed ratios whatever the total current is. The global current source is based on a 10-bits resolution DACC, and the ratios can be programmed with a 4-bit resolution. As the DACC is controlled by the micromachine with a resolution of 8 bits, we used the 2 remaining bits to further improve the accuracy of the stimulator. We thus divided the relative error over the whole scale by a factor of 3 leading this error below 5% (Figure 7.6). It is able to source and sink up to 5 mA under a 16 V DC voltage, on capacitive load. The pulse width is externally controlled; the ASIC is able to manage pulse widths as short as 500 ns, but a step of 1 μs is first proposed according to the needs.

7.2.2.2 Prototyping of the stimulation controller

Four modules constitute the stimulation controller, described in Figure 7.1. Modules are specified by means of Petri nets (PNs) based components and analyzed. Their implementation on programmable electronic device (FPGA) is done using HILECOP tool[1] (Leroux et al., 2015). For illustration, the PN model of the monitoring module is shown in Figure 7.7 (the PN models of the other modules being too complex to be shown).

Once modules have been composed and this composition analyzed, the full digital architecture is automatically translated in VHDL [c1][2] using HILECOP; the global PN, resulting from the assembly of the PNs of the modules, contains about 700 places and 800 transitions. This digital architecture constitutes a GALS system (globally asynchronous locally synchronous) since its components have their own clock and are interconnected by means of asynchronous signals. Two memories, directly described in VHDL (i.e., not generated by HILECOP), are used: the first memory is used to store the MPs, and the second one is used to store stimulation parameters that are on-line modified (remote stimulation modulation). Then this digital architecture is validated by simulation at RTL [c2][3] level using industrial tools,

[1] High-level hardware component programming tool, ensuring automatically the model transformation from PN-components to VHDL, language from which a FPGA can be programmed.

[2] Very high speed integrated circuit hardware description language.

[3] Register transfer level.

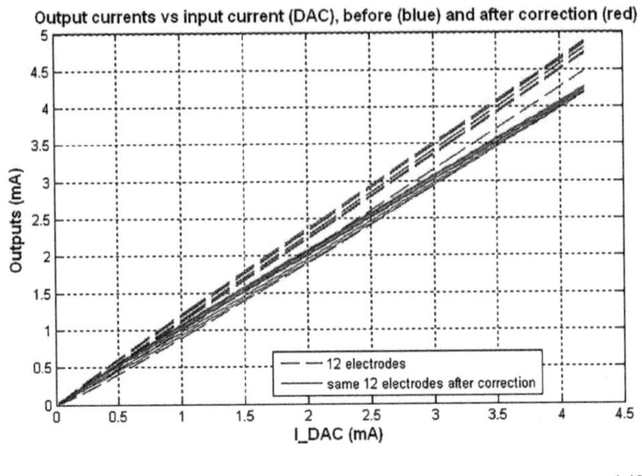

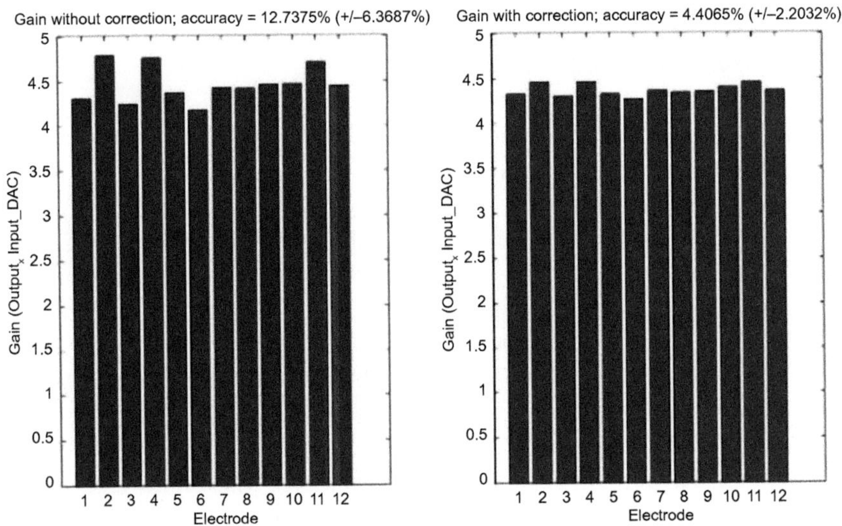

Figure 7.6 As the DACC is on 12 bits but only 8 bits are finally coded, the 4 lower bits are used to compensate the current error following an affine linear law. It cuts down the error from about 20% to less than 5% error over the full scale.

directly implemented on a programmable electronic component (FPGA) and functionally validated according to a set of validation-dedicated scenarios.

The stimulator controller embeds two memory zones to store two MPs (the configuration of the VE being included in the MP). The stimulator reacts to the following commands: *Start* (starting the stimulation according to the selected MP, i.e., *start_z0* or *start_z1*), *Stop* (stopping the ongoing

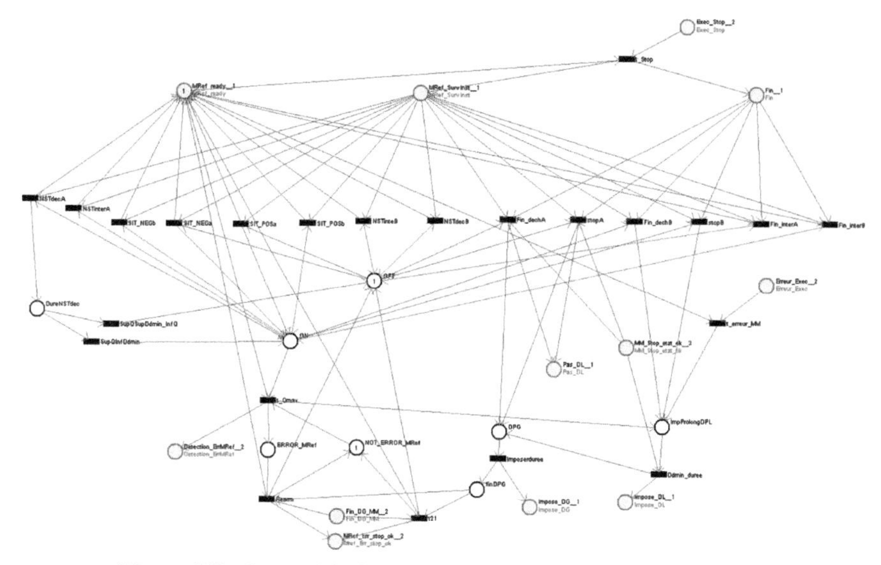

Figure 7.7 PN model of the monitoring module (reference model).

stimulation), *Rearm* (rearming the stimulation after an error detection), and *Commute* (changing the active MP). It also reacts to commands used to change values of stimulation instructions (intensity, duration, etc.).

7.2.2.3 Prototypes of stim'ND

First of all, the bench-top stimulator (Figure 7.8, left) integrates safety aspects in terms of power supply and connectivity. The power supply of Stim'ND is based on batteries whose recharge system ensures that no stimulation can be performed while batteries are recharging. It terms of connectivity, Stim'ND is based on an optically insulated USB link. Moreover, to avoid any DC component all output channels (to electrode contacts) are capacitively coupled. Last, Stim'ND embeds a safety device allowing, if required, isolating the

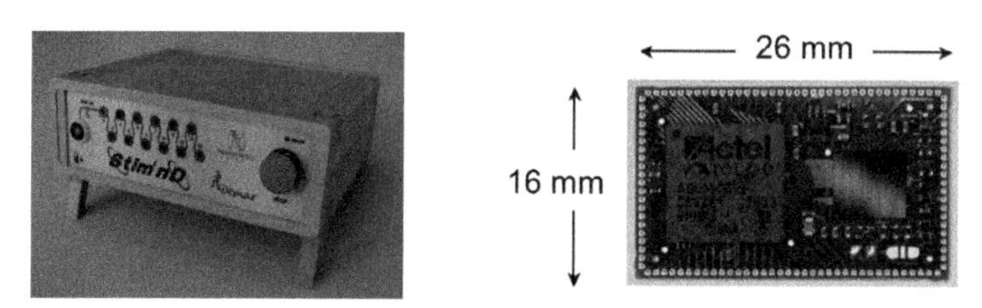

Figure 7.8 Stim'ND prototypes (left) the benchtop version (right) the miniaturized version.

patient and the stimulator outputs; this full emergency insulation of the patient from the stimulator relies on a global switch that opens all the circuits between electrodes and the stimulator, and makes a short circuit between all the electrodes to ensure no floating differential voltages. Besides, at all the software levels from the embedded digital controller (the reference models) to the remote configuration and control application, safety is considered to ensure robust and redundant checking.

7.3 Software Suite

Several software tools have been developed during the project. The first software is SENIS Manager: it allows configuring, programming, and controlling all stimulators that are connected to the controller-like computer. The second software, based on SENIS Manager API, has been developed for impedance measurement purpose in order to verify the condition of the electrode and the quality of the contact between the electrode and the neural tissue. The third software is the psychoplatform described Chapter 9. These three software applications remotely control the stimulator thanks to the embedded protocol stack that makes any stimulator a communicating unit.

To allow for such remote control of the stimulators from any of those software, the architecture has been designed and developed according to a multitier architecture. It exploits a "runtime module" that acts as a middleware in charge of communicating with stimulators (Figure 7.9); an application programming interface is provided to connect the different software to this runtime module.

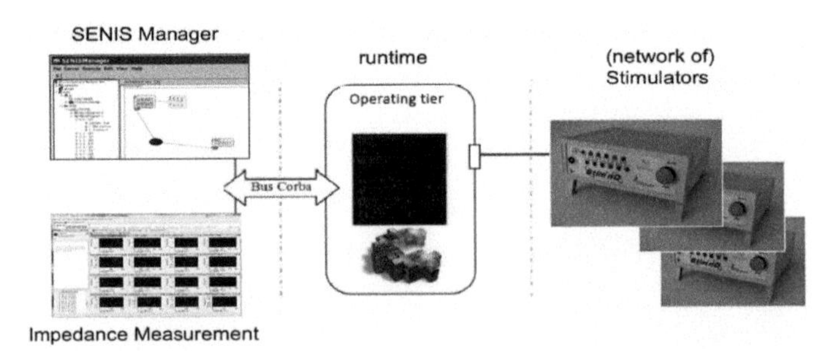

Figure 7.9 N-tier architecture allowing remote control of the stimulators.

7.3.1 SENIS Manager

SENIS Manager (Passama et al., 2011) is a software environment allowing for FES architecture configuration. It consists in configuring each stimulator that is connected to the dedicated network mastered by this controller-like computer (i.e., that running SENIS Manager). The network can be a bus or serial link. Discovering connected stimulators is based on a dedicated service that checks if any stimulator is present (64 possible nodes).

Figure 7.10 (left) shows a simple architecture composed of the PC with its associated control-box and one stimulator connected to the computer through an isolated USB link in this case.

- Functionalities associated these entities are given in Figure 7.10 (right): On the PC node the *AutoChecker* is used to periodically check (at runtime) that each unit is still connected and the *RTStimController* is used to control the stimulation process at runtime. On the stimulator node, the *Stim'ND* functionality corresponds to stimulation facilities of this kind of stimulator.
- Stimulation profile editing (Figure 7.11). It permits a graphical specification of the stimulation profile, from which it generates the corresponding MP. Depending on the instruction used some parameters of the profile can be modulated.
- Electrode configurations (Figure 7.12, left). It allows describing configuration of the 12 channels (poles), setting active poles, their polarity and ratios of current repartition between active poles knowing that the sum of current ratios for cathodes must be 100% (as for anodes).
- Stimulators programming consists in downloading MPs in its memory areas as well as setting the parameters of its reference model (safety

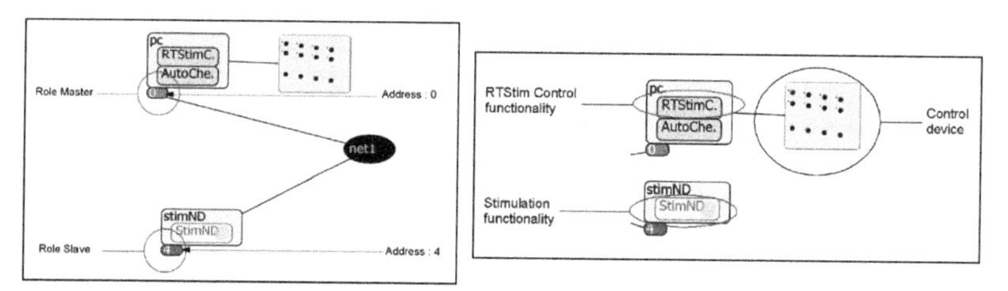

Figure 7.10 Example of simple architecture (left). Functionalities associated to entities (right).

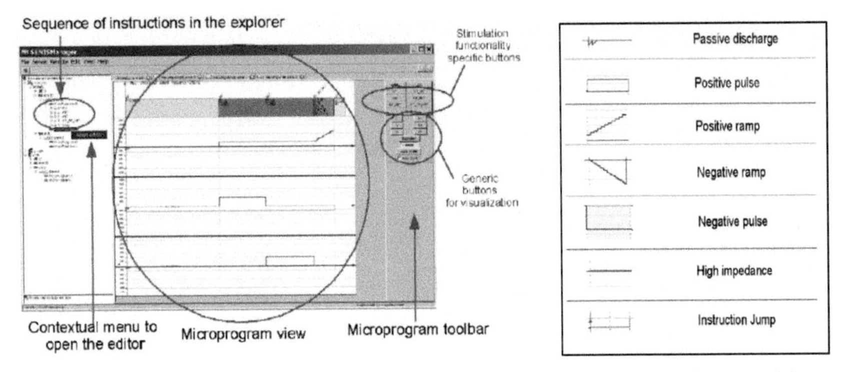

Figure 7.11 Stimulation profile editing (left). Correspondence between icons and instructions (right).

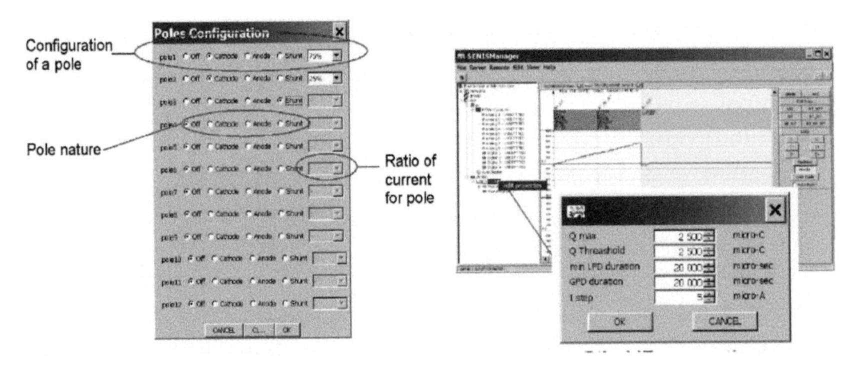

Figure 7.12 Electrode configuration (left). Configuring reference model parameters (right).

module), as, for instance, the maximum quantity of injected charge (Figure 7.12, right).

- Once configured and programmed, any stimulator can be remotely controlled. Stimulation control relies on two main facilities that are commands and stimulation modulations through online modification of the stimulation parameters. This can be done through the HMI as well as from a dedicated control-box (Figure 7.13) with four push-buttons and eight rotary ones; association of push-buttons to *RT commands* (start, stop, commute, rearm) and rotary ones to *RT updates* of parameters (*I*, *PW*) is simply done through the software. For each rotary button, the user has to associate: the concerned entity (i.e., stimulator), the concerned MP, the RT instruction and the parameter to be modulated, as well as min and max values for the given parameter (i.e., min and

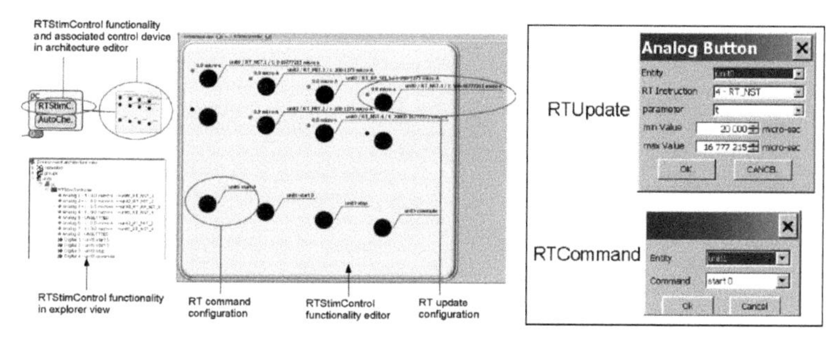

Figure 7.13 Configuring the control-box (left). Configuration of buttons (right).

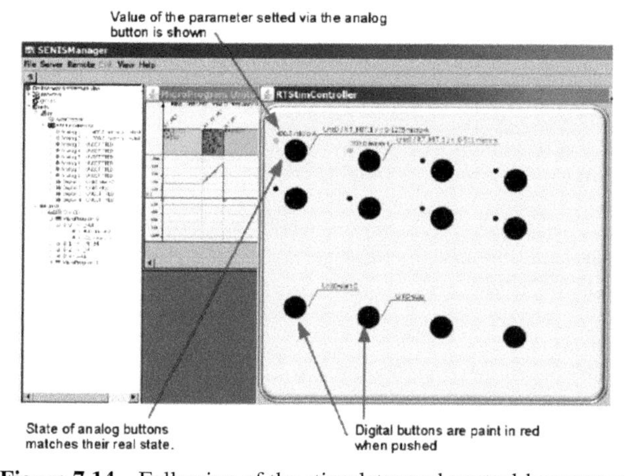

Figure 7.14 Following of the stimulator and control-box states.

max values of the button). Using a control-box allows the practitioner to keep attention on the stimulated target when controlling the stimulation rather than on the screen.

Once started, the stimulator state and control-box parameters are continuously monitored as shown Figure 7.14 where the stimulation is running and parameters are modulated, and Figure 7.15 where a constraint violation has been detected by the stimulator that immediately went to a safe state, waiting to be rearmed.

Different stimulations can be specified playing with the MPs' internal repetition and computer-driven sequencing of MP execution (Figure 7.16).

SENIS Manager can deal with individual entities (unicast addressing) and groups of entities (multicast addressing), belonging to the same network. Thus it allows for synchronized remote control of stimulators (RT commands only) for multisite coordinated stimulation.

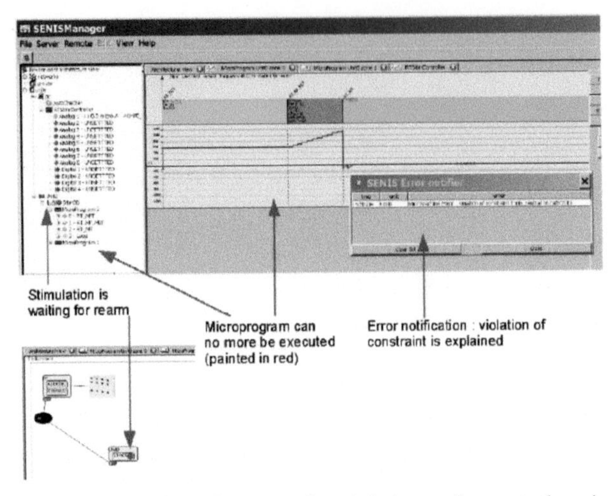

Figure 7.15 Notification of a constraint violation to the control environment.

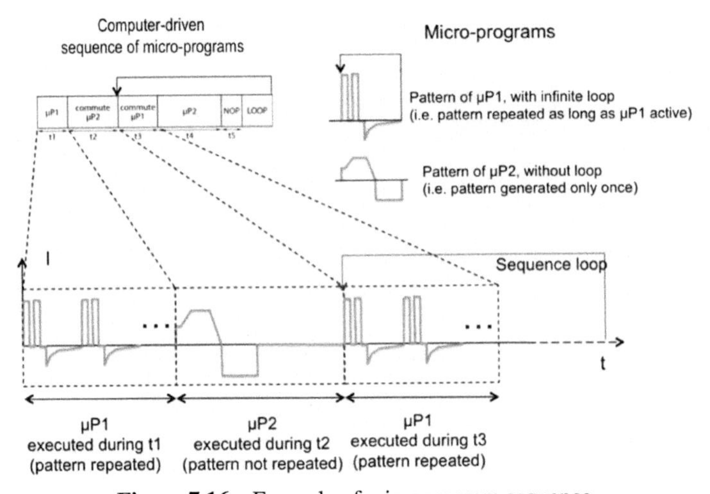

Figure 7.16 Example of microprogram sequence.

7.3.2 Impedance Follow-up Software

In order to verify the condition of the electrode and the quality of the contact between the electrode and the neural tissue, during the implantation of the TIME electrode and at few time instants after the implantation, the impedance of each site of the four implanted TIME electrodes has been measured using indirect method. For that purpose a small controlled-current stimuli (biphasic rectangular pulses of 40 μA, 300 μs duration of each phase) were generated using each of the TIME electrode stimulation sites as a cathode and the reference (GND) site of each electrode as the anode. While the stimuli were generated the voltage drop between the cathode and the reference site as well as the voltage drop on the 1 kΩ resistor inserted in series with the TIME electrode have been measured. Based on those two measurements it was possible to determine the impedance of each electrode site versus each of the two reference sites (left and right) of the same electrode as well as between the two reference sites of each electrode.

In order to facilitate those measurements and to ensure that the measurements during follow-ups are performed in the same way, two applications have been developed. The first one has been developed using the Microsoft Visual Basic Express programing language (Figure 7.1) and allowed to generate a sequence of stimuli for various sites of the TIME electrode using the Stim'ND stimulator. It also allowed pausing stimulation after generation of stimuli for the particular site, leaving the time for the experimenter to evaluate the results. The second application was developed using the National Instruments LabVIEW Signal Express programing language and allowed to record the voltage drops in the moment of stimulation and also to immediately display results of the measurements for up to eight different stimulation sites combinations (Figure 7.17 – example of the results obtained right after the surgery for the TIME electrode implanted into the median nerve, results

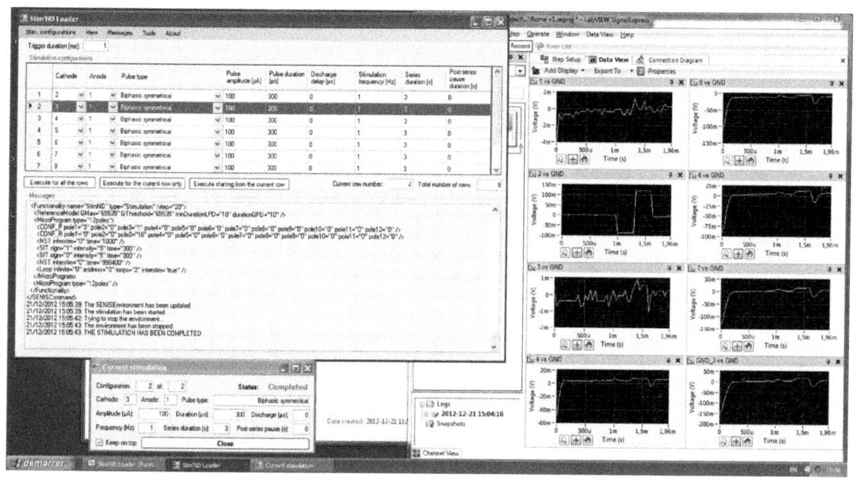

Figure 7.17 The screenshot shows the control of the stimulator on the left and the resulting current-voltage curves from which the estimation is performed (ratio of U and I at the end of the active phase).

obtained for the stimulating sites on the right side of the electrode versus the right reference site).

Each of the implanted TIME electrodes had 16 contacts: seven on the left side, seven on the right side, and two reference sites (one on the left side – LGND and one on the right side – RGND). Therefore, for each TIME electrode four measurement series were performed:

- Seven contacts on the left site and the right reference site versus the left reference site
- Seven contacts on the left site and the left reference site versus the right reference site
- Seven contacts on the right site and the left reference site versus the right reference site
- Seven contacts on the right site and the right reference site versus the left reference site

The measurements for each site has been performed versus each of the reference sites in order to determine the reference site with lower impedance and to be able to estimate impedance of all the stimulating sites even if one of the reference sites was broken.

The developed application allowed performing each measurement series (i.e., measurements for eight combinations of stimulating sites) in less than

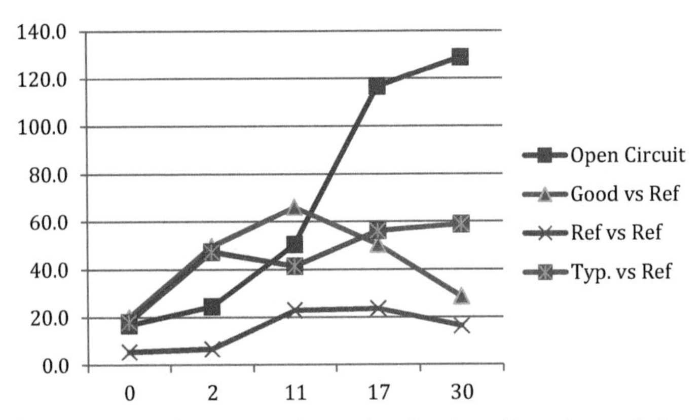

Figure 7.18 The four graphs represent the rough estimation of impedances (kohms) of all the 56 contacts. Two different profiles of impedances were found (green increase then decline, red constant increase). The references are much bigger so the impedance is much lower. Finally, open circuits have a clear and strong increase of the impedance from day 17.

1 min. Because for each of the four TIME electrodes four measurement series were performed (4 measurement series $\times$ 4 TIME electrodes), using those applications it was possible to perform the measurement of impedance for all the stimulating sites of the four TIME electrodes in less than 15 min.

The impedance measurements have been performed during the implantation of the electrodes, in order to verify the correct position of the electrode and to provide feedback to the surgeon, and also right after the surgery as well as 2, 11, 17, and 30 days after the implantation in order to track the changes of the impedance during the course of the experiment. The results of the measurements are presented in Figure 7.18. It may be observed that at the last day of measurement (i.e., 30 days after the implantation) only seven sites had higher impedance than expected (i.e., above 80 kΩ).

7.4 Discussion

The stimulator Stim'ND has shown its suitability to achieve selective multipolar stimulations. However, the different control modalities of the stimulator, through different software and operators, have also revealed some necessary improvements. Modulating the stimulation profile simultaneously playing with several parameters of several instructions still is a cumbersome and inefficient procedure. A new processor, more particularly a new set of instructions, must be designed to allow the modulation of all or part of the stimulation profile with notably preservation of charge-balanced stimulation.

As stimulation often relies on executing different kinds of stimulation waveforms on different electrode configurations, we designed the stimulator according to 3D electrode (VE) and MP concepts. Nevertheless at this stage the description of a 3D electrode is inserted into a MP, thus limiting the possible reuse of each of these entities. The stimulator should be able to be programmed via these two entities, namely the tuple {VE, MP}, constituting so key elements of a high-level programming language.

Finally, SEF applications may require the sequencing of different stimuli (i.e., different tuples {VE, MP} executed in sequence). Remotely controlling this sequencing poses a performance problem in the sequence of stimulations, the remote control necessarily inducing a latency of communication and decoding. One of the solutions would be to decentralize this functionality, i.e., to embed it directly into a stimulation unit, at the expense of complexity, surface, and consumption.

References

Andreu, D., Guiraud, D. and Souquet, G. (2009). A distributed architecture for activating the peripheral nervous system. *Journal of Neural Engineering*, 6(2):026001.

Badia, J., Boretius, T., Andreu, D., Azevedo-Coste, C., Stieglitz, T. and Navarro, X. (2011). Comparative analysis of transverse intrafascicular multichannel, longitudinal intrafascicular and multipolar cuff electrodes for the selective stimulation of fascicles. *Journal of Neural Engineering*, 8:036023.

Dali, M., Rossel, O., Andreu, O. D., Laporte, L., Hernández, A., Laforet, J., Marijon, E., Hagège, A., Clerc, M., Henry, C. and Guiraud, D. (2018). Model based optimal multipolar stimulation without *a priori* knowledge of nerve structure: application to vagus nerve stimulation. *Journal of Neural Engineering*, 15(4):046018.

Davis, R., Houdayer, T., Andrews, B., Emmons, S. and Patrick, J. (1997). Paraplegia: prolonged closed-loop standing with implanted nucleus FES-22 stimulator and Andrews' foot-ankle orthosis. *Stereotactic and Functional Neurosurgery*, 69:281–287.

Djourno, A. and Eyries, C. (1957). Prosthèse auditive par excitation électrique à distance du nerf sensoriel à l'aide d'un bobinage inclus à demeure [auditory prosthesis for electrical excitation at a distance from a sensory nerve with the help of an embedded electrical coil]. Presse Médicale, 35:14–17.

Godary-Dejean, K. and Andreu, D. (2013). Formal validation of a deterministic MAC protocol. *ACM Transactions on Embedded Computing Systems*, 12, 6:1–23.

Guiraud, D. (2011). Interfacing the neural system to restore deficient functions: from theoretical studies to neuroprothesis design, Comptes Rendus Biologies, 335:1–8.

Guiraud, D., Andreu, D., Bernard, S., Bertrand, Y., Cathébras, G., Galy, J. and Techer, J. D. (2010). Device for distributing power between cathodes of a multipolar electrode, in particular of an implant. US 7768151 B2.

Guiraud, D., Azevedo Coste, C., Benoussaad, M. and Fattal, C. (2014). Implanted functional electrical stimulation: case report of a paraplegic patient with complete SCI after 9 years. *Journal of NeuroEngineering and Rehabilitation*, 11:15.

Kobetic, R. Triolo, R. J. and Marsolais, E. B. (1997). Muscle selection and walking performance of multichannel FES systems for ambulation in paraplegia. *IEEE Transactions on Rehabilitation Engineering*, 5(1):23–29.

Kobetic, R., Triolo, R. J., Uhlir, J. P., Bieri, C., Wibowo, M., Polando, G., Marsolais, E. B., Davis Jr, J. A., Ferguson, K. A. and Sharma, M. (1999). Implanted functional electrical stimulation system for mobility in paraplegia: a follow-up case report. *IEEE Transactions on Rehabilitation Engineering*, 7(4):390–398.

Kralj, A. and Bajd, T. (1989). Functional Electrical Stimulation: Standing and Walking After Spinal Cord Injury. Boca Raton: CRC Press Inc.

Leroux, H., Andreu, D. and Godary-Dejean, K. (2015). Handling exceptions in Petri nets based digital architecture: from formalism to implementation on FPGAs. *IEEE Transactions on Industrial Informatics*, 11, 4:897–906.

Liberson, W. T., Holmquest, H. J., Scot, D. and Dow, M. (1961). Functional electrotherapy: stimulation of the peroneal nerve synchronized with the swing phase of the gait of hemiplegic patients. *Archives of Physical Medicine and Rehabilitation*, 42:101–105.

Mastinu, E., Doguet, P., Botquin, Y., Håkansson, B. and Ortiz-Catalan, M. (2017). Embedded system for prosthetic control using implanted neuromuscular interfaces accessed via an osseointegrated implant. *IEEE Transactions on Biomedical Circuits and Systems*, 11(4):867–877.

Passama, R., Andreu, D. and Guiraud, D. (2011). Computer-based remote programming and control of stimulation units. 5th IEEE/EMBS International Conference on Neural Engineering, Cancun, Mexico.

Rijkhoff, N. J. M. (2004). Neuroprostheses to treat neurogenic bladder dysfunction: current status and future perspectives. *Childs Nervous System ChN Social Journal of the International Society for Pediatric Neurosurgery*, 20(2):75–86.

Schiefer, M. A., Triolo, R. J. and Tyler, D. J. (2008). A model of selective activation of the femoral nerve with a flat interface nerve electrode for a lower extremity neuroprosthesis. *IEEE Transactions on Neural Systems and Rehabilitation Engineering*, 16:195–204.

Smith, B., Zhengnian, T., Johnson, M. W., Pourmehdi, S., Gazdik, M. M., Buckett, J. R. and Peckham, P. H. (1998). An externally powered, multichannel, implantable stimulator-telemeter for control of paralyzed muscle. *IEEE Transactions on Biomedical Engineering*, 45(4):463–475.

8

Computerized "Psychophysical Testing Platform" to Control and Evaluate Multichannel Electrical Stimulation-Based Sensory Feedback

Bo Geng[1], Ken Yoshida[2], David Guiraud[3], David Andreu[3], Jean-Louis Divoux[4] and Winnie Jensen[1,*]

[1]Department of Health Science and Technology, Aalborg University, Denmark
[2]Department of Biomedical Engineering, Indiana University – Purdue University Indianapolis
[3]LIRMM, University of Montpellier, INRIA, CNRS, Montpellier, France
[4]Axonic, MXM, France
E-mail: wj@hst.aau.dk
*Corresponding Author

8.1 Introduction

Use of electrical stimulation through implanted neural interfaces makes it possible to selectively activate afferent neurons and provide natural sensory feedback. However, effective stimulation patterns for creating natural sensory feedback are not uniquely characterized in the literature. Thus, there is often a need to evaluate various combinations of different stimulation parameters for specific applications, which can produce a wide range of possible stimulation patterns in a multichannel stimulation system. It makes the use of sensory feedback an impracticable and time-consuming task. We therefore designed and implemented a computerized tool referred to as a "Psychophysical Testing Platform" to easily control multichannel stimulation and characterize the evoked sensations. The tool was tested in a clinical trial including one amputee with the aim to relieve his phantom limb pain (PLP) by manipulation of phantom sensation using intraneural stimulation. The tool may also be utilized in systems based on surface electrical stimulation.

8.2 Sensory Feedback

Amputation of a limb involves the complete transection of afferent and efferent nerves innervated the removed limb. Sensory feedback from the missing part of the limb is thus severely impaired. While most amputees experience that the absent limbs still exist (Kooijman et al., 2000), artificial activation of the residual afferent neurons likely enhances the sensory feedback by creating more specific sensations, e.g., joint position or finger movement. The enhanced sensory feedback may be utilized in prosthetic hand control or PLP treatment (see, e.g., Flor et al., 2001; Rossini et al., 2010; Dietrich et al., 2012).

Electrical stimulation has been recognized as one of the feasible approaches to artificially activate sensory neurons, which can, according to where the stimulation is applied, broadly be grouped into two types: cutaneous stimulation and direct nerve stimulation (Riso, 1999). Cutaneous stimulation may substitute impaired sensibilities by accessing the tactile senses in the skin (Szeto and Saunders, 1982). Users are typically provided with coded stimuli (such as modulation of pulse rate) and learn to relate these codes to specific sensory information, e.g., pinch force (Shannon, 1979). However, intensive cognitive load is usually required for a user to correctly interpret the coded signal (Prior et al., 1976).

Stimulation directly applied to the peripheral nerves through implanted nerve-electrode interfaces makes it possible to generate natural sensations and provide more intuitive sensory feedback (Anani and Korner, 1979; Dhillon et al., 2005). It has been demonstrated that intraneural stimulation via implanted microelectrodes was capable of evoking sensation of touch, joint position, and movement referred to the amputated limb (Dhillon et al., 2005; Rossini et al., 2010). The orientation of the stimulation sites of the implanted electrodes is usually blinded due to complex histological characteristics of peripheral nerves as well as due to some limitations of the current surgical procedures (Stewart, 2003; Harreby et al., 2012; Kundu et al., 2012). As previously mentioned, the stimulation patterns capable of producing natural sensory feedback are not well documented in the literature. Therefore, there is a need to consider a variety of possible combinations of stimulation parameters that may generate sensations such as touch/pressure, finger movement, and vibration (Dhillon and Horch, 2005) or tingle, touch, vibration, buzz, pinch in cutaneous stimulation (Kaczmarek et al., 1991; Pfeiffer, 1968).

The magnitude of a sensation is commonly estimated using scaling methods by assigning a number to the perceptual event such as sensation (Stevens, 1957). A 10-point visual analogue scale (VAS) is widely used, with 0 representing "no sensation" and 10 "the upper limit of a sensation or pain." There are also a number of other types of linear scales such as Likert scale and Borg scale. A particular scaling method may outperform the others in specific circumstances (Grant et al., 1999).

8.3 Sensory Feedback for Phantom Limb Pain Treatment

In 50–80% of amputees, pain occurs in the missing limb, known as PLP (Weeks et al., 2010). There are no effective, long-lasting treatments currently available for PLP and it has not been completely understood why and how the phantom pain develops. However, cortical reorganization has been discussed as a plausible cause for the development of PLP (Ramachandran et al., 1992; Flor et al., 1995) and a positive relation was found between the amount of plasticity in the primary somatosensory cortex (S1) and the severity of PLP (Florence et al., 2000; Lotze et al., 2001; Karl et al., 2004).

Several studies demonstrated a positive effect of enhancing sensory feedback on reversal of cortical changes and relief of PLP. For instance, the patients who received daily training in sensory discrimination of surface electrical stimuli applied to the stump, experienced reduction of PLP after 2 weeks (Flor et al., 2001). In another study, training in control of a robotic hand with limited amount of sensory feedback significantly reduced PLP in a human amputee implanted with four intrafascicular electrodes in the nerve stump. The reduction in PLP lasted several weeks after removal of the electrodes and changes in sensorimotor cortex topography were shown (Rossini et al., 2010). A recent study found that use of a prosthesis that provides somatosensory feedback on the grip strength effectively alleviated PLP (Dietrich et al., 2012). The evidence suggested the likelihood to suppress PLP by providing intensive, natural sensory feedback to amputee patients.

8.4 Psychophysical Testing Platform Design Strategy and Principles

The psychophysical testing platform was designed as a part of the "TIME prototype system" to efficiently test, deliver, and evaluate generated sensory feedback (see introduction).

The aim of the TIME (transverse intrafascicular multichannel electrodes) project was to develop an implantable neural prosthesis system with sufficient stimulation selectivity to generate phantom sensations and explore the possibility of using the method as a potentially effective treatment for PLP. The project hypothesized that manipulating phantom sensations using selective stimulation of the nerve stump may reverse cortical reorganization and consequently mitigate PLP.

TIME electrodes, each consisting of 12 stimulation sites, were designed and manufactured by IMTEK (University of Freiburg, Germany) based on micromachining technologies on flexible polymeric substrates (Boretius et al., 2010) were used as the interface for the peripheral nerves, see also Chapter 3. A customized 12-channel stimulator was developed (Montpellier Laboratory of Informatics, Robotics, and Microelectronics and MXM Neuromedics, France) to deliver the electrical stimulation, see also Chapter 7.

The psychophysical testing platform consisted of two computers interacting with each other – i.e., subject performs psychophysical tests for characterization of the delivered sensory feedback on Computer #1 and the experimenter controls and monitors the stimulation process on computer #2.

Definition of the system functionalities was based on the experimental tasks to be conducted. Three main experimental tasks were identified: (1) determination of threshold, (2) characterization of sensation, and (3) repeated application of "useful" stimulation.

- *Experimental Task 1: Determination of the sensation threshold and upper limit of sensation.* Before applying stimulation for pain relief treatment, it is necessary to determine which of the TIME electrode active sites are functional and how much electricity should be injected into each active site to elicit natural sensations referred to the phantom hand. Therefore, the sensation threshold and the upper limit of a sensation first needs to be determined. The sensation threshold is defined as the current level where subject can just barely detect that a stimulus is delivered. The upper limit of a sensation is defined as the current level where the nature or the location of the sensation changes, or when the sensation becomes uncomfortable or painful. To determine the sensation threshold, the intensity of the stimuli presented to the subject should initially be sufficiently weak and then increase in successive steps. Different psychophysical methods can be adopted to identify the sensation threshold and the upper limit of the sensation (Ehrenstein et al., 1999).

- *Experimental Task 2: Sensation characterization.* After thresholds are determined, the perceived location, type, and intensity of the sensation should be characterized for those active sties that were defined as functional in Experimental Task 1. This task is important to identify the stimulation patterns that can produce natural phantom sensations which are interpreted as meaningful by the subject. The range of current between the sensation threshold and the upper limit of sensation are applied. Other parameters that are interested include: pulse duration, pulse frequency, single-pulse vs. pulse train, number of pulses, monopolar vs. bipolar, as well as stimulation site combinations.
- *Experimental Task 3: Repeated application of natural sensory feedback.* Repeated application of the stimulation patterns that were identified as "meaningful" in Task 2 may reinforce the effect of sensory feedback on cortical plasticity and PLP. A subset of optimal stimulation patterns selected based on the results in the Task 2 is repeatedly applied in the experimental task.

Four main functionalities were defined according to the tasks above described.

- *Functionality 1: Decide stimulation parameters.* An interface is needed for the experimenter to configure the stimulation parameters to be varied. Each stimulus usually needs to be repeated multiple times and randomized to obtain a data set that later can be statistically analyzed. The interface should thus allow repeated configuration of one or a series of stimuli, as well as randomization of stimuli.
- *Functionality 2: Determine thresholds.* Threshold is usually determined by delivering a set of stimuli with an intensity in the vicinity of the threshold. An interface is developed for the subject to indicate whether or not the stimulus just delivered was perceived or perceived as the upper limit. The subject's response and corresponding stimulation parameters need to be tracked for later calculation of threshold.
- *Functionality 3: Characterize sensations.* A sensation can be characterized by three main attributes: location, type, and intensity. An interface is needed for this psychophysical testing, which comprises a questionnaire

presented to the subject for evaluating the three sensation attributes. The subject's response and corresponding stimulation parameters is tracked for later identification of optimal stimulation patterns, or analysis of the relation between a parameter and evoked sensation.

- *Functionality 4: Automated repetition of stimulation and characterization.* The large amount of stimulus patterns to be investigated to determine the thresholds or characterize the sensations requires an efficient way to deliver stimuli and characterize the evoked sensation. An automated process of stimulus delivering, sensation characterization, and data collection is the solution to address the issue. The system supports the automated process, in which each session proceeds as illustrated in Figure 8.2.

8.5 Software Components

The software was developed in the LabVIEW environment (National Instruments). Control of the electrical stimulator was implemented through accessing a set of application programming interface (API) functions built in dynamic linked libraries provided by the stimulator developer.

The specified functionalities were implemented in two subsystems, i.e., the stimulator and experiment control (SEC) subsystem (located in Computer #1, see Figure 8.1) and the interactive subject interface (ISI) subsystem (located in Computer #2, see Figure 8.2).

Stimulator and experiment control (SEC): The SEC subsystem provides a tool for the experimenter to configure stimulation parameters, and monitor

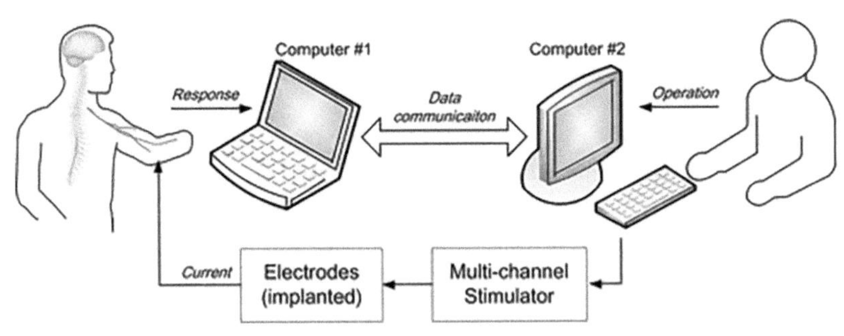

Figure 8.1 Prototype system in the TIME project. The psychophysical testing platform was implemented on Computer #1 and Computer #2 to interact with the experimenter and the subject.

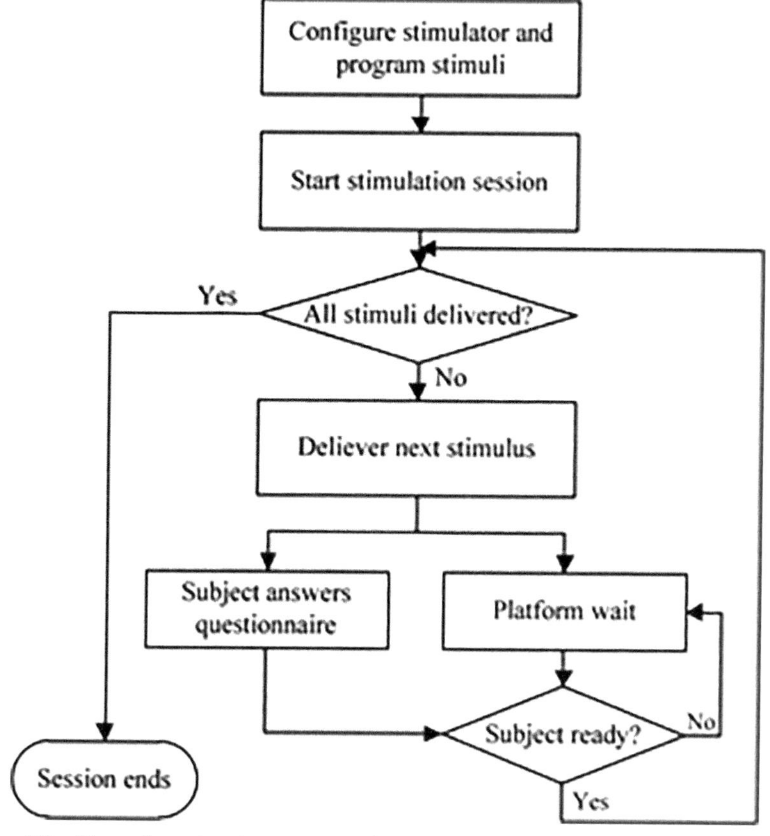

Figure 8.2 Flow chart showing automated process of stimulus delivering and sensation measurement in one stimulation session.

stimulation process and experimental progress. To secure safe delivery of the electrical stimulation it is possible for the experimenter to override the otherwise automated stimulation process.

Interactive subject interface (ISI): The ISI subsystem provides the interface to perform psychophysical testing with the subject, i.e., threshold determination and sensation characterization. This subsystem also collects all the subject's responses for later data analysis.

The functionalities defined for the SEC were implemented in five modules physically grouped in the main graphical user interface (GUI), as shown in the screenshot (Figure 8.3). The five modules are described in details as follow.

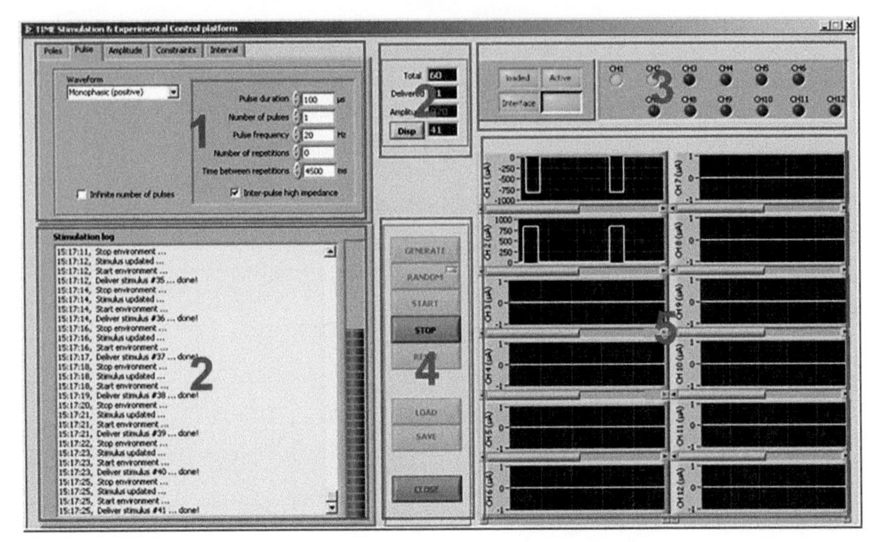

Figure 8.3 Screenshot of the main GUI of the SEC software with the five modules identified (i.e., the module numbers are shown in the center of each module box).

Table 8.1 A list of stimulation parameters implemented in SEC software

Parameter	Range	Step Size
Pulse waveform	Monophasic (negative or positive)	N/A
	Biphasic (symmetric, positive following negative)	
	Biphasic (symmetric, negative following positive)	
	Biphasic (arbitrary amplitude and pulse duration)	
Amplitude	5 mA	20 μs
Pulse duration	500 μs	1 μs
Number of pulses	1–5000 pulses	1
Frequency	1–1000 Hz	1 Hz

Module 1: Stimulation parameter configuration. In this module different parameters can be configured in a panel consisting of several pages. Table 8.1 is a summary of the range and step size of each stimulation parameter. The panel also includes a page where the stimulation constraints can be configured to ensure the safety. If a stimulus violates the constraints, the stimulation session will stop.

Module 2: Progress monitoring. A progress bar is used to indicate how far a stimulation session has progressed. A textual indicator (light blue box) shows specific status information of the stimulation session (e.g., which stimulus was just delivered, waiting for a response from the subject). This information is automatically saved in a log file. In addition, four numeric indicators are used to respectively display: (1) the total number of stimuli in the ongoing stimulation session, (2) the number of stimuli already delivered, (3) the current level of the stimulus just delivered, and (4) the index of the stimulus just delivered.

Module 3: Stimulator status monitoring. Four squared "LED" indicators are used to indicate the status of interaction between the software and the stimulator. These four indicators are used for trouble-shooting if the stimulation process accidently stops or hangs during communicating with the stimulator. Twelve round-shape "LED" indicators are used to indicate active cathode channels (i.e., light up when active).

Module 4: Experimental control. This module allows to control the experiment and choose the way that the configured stimuli are delivered. The commands include:

- Add one or a series of increasing stimuli with constant step size.
- Randomize the order of the stimuli to be delivered.
- Start/stop/continue a stimulation session.
- Save the stimulus configurations to an external text file.
- Load stimulus configurations from a previously saved file.
- Clear and reset the stimuli.

Module 5: Graphical display. This module provides a graphical display of the stimulus waveform just delivered in the 12 stimulator output channels.

8.6 Implementation of ISI Subsystem

The ISI subsystem software includes two main user interfaces: the interface for threshold determination and the interface for sensation characterization. The interface for threshold determination provides the subjects with a YES or NO choice on whether or not the stimulation was perceived, or whether or not the stimulation was perceived as the upper limit of a sensation (Figure 8.4).

In the interface for sensation characterization, three questions are implemented with regard to the sensation type, location, and magnitude, respectively (see Figure 8.5).

Question 1: Please choose one or more words to describe the sensation you felt. A list of words describing possible evoked sensation types is predefined based on previous studies (Dhillon et al., 2005; Dhillon and Horch, 2005; Rossini et al., 2010). The list includes touch/pressure, vibration, tugging, spider crawling, pinch, pain, wrist flexion, wrist extension finger flexion, finger extension, cold and warm. One or more types can be selected.

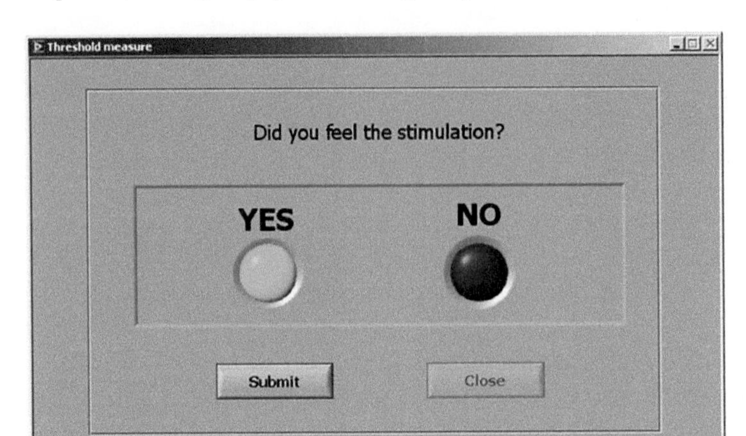

Figure 8.4 Screenshot of user interface for threshold determination.

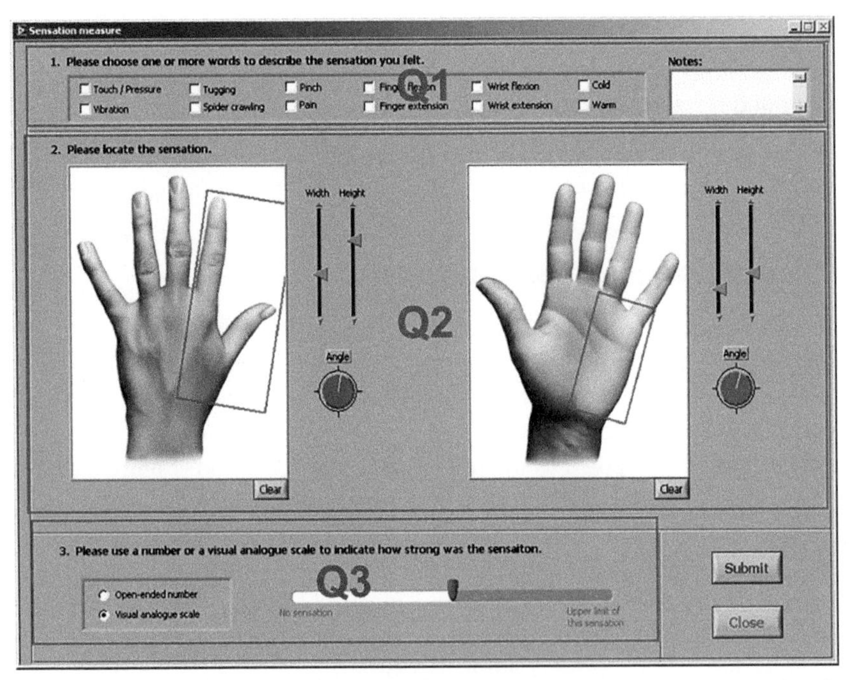

Figure 8.5 Screenshot of user interface for characterization of the sensation type, location, and magnitude, each corresponding to a question in the red box.

In case the perceived sensation is not covered by the list, an option of making a note is provided to the subject.

Question 2: Please locate the sensation. Two pictures of the front and back view of a human hand are provided for the subject to locate the evoked sensation. The subject needs to point the cursor to the center of the perceived region using the mouse. A red-color rectangular box will then be marked on the hand picture. The width, length and angle of the box can be adjusted.

Question 3: Please use a number or a visual analogue scale to indicate how strong you felt the sensation. Two approaches of measuring the sensation strength were implemented: visual analogue scale (VAS) and open-ended number. A VAS is a horizontal bar anchored by word descriptors at each end (left end: no sensation, right end: upper limit of a sensation). The subject can mark a point by moving the slider, to indicate how strong the sensation is. The VAS score is then determined by the distance from the left end of the bar to the point that the subject marks. Alternatively, the subject can use an open-end number to indicate the magnitude of a sensation.

8.7 Communication Between SEC and ISI

The SEC and ISI software are able to communicate and exchange data through a local area network to achieve the automated process of stimulation and evaluation. The physical substrate of the communication is an Ethernet crossover cable, which directly connects the two computers allowing data transfer across the network. The LabVIEW DataSocket has been used to realize data communication between two applications residing in two computers.

8.8 Use of the Psychophysical Testing Platform

Training of amputee subject. One amputee volunteer was recruited for the clinical test of the TIME prototype system (see also Chapter 9). Before the subject had the electrodes implanted he went through a training session to familiarize him with the automated stimulation-characterization procedure, as well as the interactive interfaces for psychophysical testing. Comparing the use of paper-based psychophysical questionnaire, the subject reported that the system assisted and promoted the process of the psychophysical testing, which helped him concentrate more on the actual experiments.

Evaluation by clinical doctors. A questionnaire was used to evaluate the usability of the software and user satisfaction of the interfaces. It was developed based on a modification of IBM The Post-Study System Usability Questionnaire (Lewis, 1995) and the Questionnaire for User Interface Satisfaction (Chin et al., 1988). Only relevant questions from the original questionnaires were included in our evaluation. This questionnaire was given to our clinical partner who was using the software in clinical experiments. The response to the questionnaire helped us understand what aspects of the software they were particularly concerned about and satisfied with.

Use in clinical trials. When the electrodes were implanted in two main nerves in the subject's forearm (i.e., the median and ulnar nerve), the software part of the TIME prototype system was used for clinical tests. SEC was used to define specific stimulation sequences, define, and control the delivery of the electrical stimulation, and monitor the progress of the experiment. ISI was used to perform psychophysical testing, where the subject filled in questionnaires to describe and quantify the perceived sensation referred in the phantom hand (see Chapter 9).

8.9 Discussion

Following amputation, the complete truncation of afferent and efferent nerves leads to lack of sensory feedback from the missing limb, and this can produce a phantom pain effect. Artificial activation of sensory neurons may then be considered as a solution to recover the impaired sensibility. Microfabricated neural interfaces enable direct nerve stimulation with high selectivity. However, morphological complexity and limited knowledge about optimal stimulation strategies make it necessary to investigate a large number of different combinations of stimulation parameters. To overcome these limitations, a computerized tool may be used to efficiently evaluate the sensory feedback in a multichannel, intraneural stimulation setting as described here.

We sought to evaluate the usability and user satisfaction of the software. However, the evaluation was limited due to the simple fact that the system was only used with one amputee subject.

Based on the experience obtained from the clinical tests, we believe that the psychophysical testing platform may be improved in the following aspects. For instance, a resizeable, rotatable rectangular box was used to localize a sensation, since the perceived region could be in an irregular shape. An improved tool should instead allow arbitrary drawing to localize the sensation more precisely and speed up the process further. Moreover, bilateral upper-extremity amputees may not be able to control a computer mouse conveniently. Using a touch screen could be easier, especially for above-elbow amputees.

The platform was designed to collect the data from threshold and sensation measure experiments. However, to counteract cortical plasticity and consequently relieve PLP, it is necessary to carry out repeated, intensive stimulation sessions. Therefore, a set of optimal stimulation patterns should be identified based on the results in sensation characterization. The optimal stimulation patterns is defined as those capable of eliciting clear, meaningful, distinct sensation referred to the phantom hand, e.g., finger movement, touch, joint position, etc. As such, a module that can automatically select optimal stimulation patterns can be considered in future development.

References

Ehrenstein, W. H. and Ehrenstein, A. (1999). Psychophysical methods. In Modern techniques in neuroscience research. Berlin: Springer.

Gescheider, G. A. (1997). Classical psychophysical theory. Psychophysics: the fundamentals (3rd ed.). Lawrence Erlbaum Associates.

Marks, L. E. and Gescheider, G. A. (2002). Psychophysical scaling. Stevens' handbook of experimental psychology.

Anani, A. and Korner, L. (1979). Discrimination of phantom hand sensations elicited by afferent electrical nerve stimulation in below-elbow amputees. Med Prog Technol 6:131–135.

Boretius, T., Badia, J., Pascual-Font, A., Schuettler, M., Navarro, X., Yoshida, K. and Stieglitz, T. (2010). A transverse intrafascicular multichannel electrode (TIME) to interface with the peripheral nerve. Biosens Bioelectron 26:62–69.

Chin, J. P., Diehl, V. A. and Norman, K. L. (1988). Development of an instrument measuring user satisfaction of the human-computer interface. Proceedings of the SIGCHI conference on human factors in computing systems, pp. 213–218, ACM/SIGCHI.

Dhillon, G. S. and Horch, K. W. (2005). Direct neural sensory feedback and control of a prosthetic arm. IEEE Trans. Neural Syst. Rehabil. Eng. 13:468–472.

Dhillon, G.S., Krüger, T. B., Sandhu, J. S. and Horch, K. W. (2005). Effects of short-term training on sensory and motor function in severed nerves of long-term human amputees. J. Neurophysiol 93:2625–2633.

Dietrich, C., Walter-Walsh, K., Preißler, S., Hofmann, G. O., Witte, O. W., Miltner, W. H. R. and Weiss, T. (2012). Sensory feedback prosthesis reduces phantom limb pain: Proof of a principle. Neurosci. Lett. 507:97–100.

Downing, S. M. (2004). Reliability: On the reproducibility of assessment data. Med Educ 38:1006–1012.

Flor, H., Denke, C., Schaefer, M. and Grüsser, S. (2001). Effect of sensory discrimination training on cortical reorganisation and phantom limb pain. Lancet 357:1763–1764.

Flor, H., Elbert, T, Knecht, S., Wienbruch, C., Pantev, C., Birbaumer, N., Larbig, W. and Taub, E. (1995). Phantom-limb pain as a perceptual correlate of cortical reorganization following arm amputation. Nature 375:482–484.

Florence, S. L., Hackett, T. A. and Strata, F. (2000). Thalamic and cortical contributions to neural plasticity after limb amputation. J. Neurophysiol 83:3154–3159.

Grant, S., Aitchison, T., Henderson, E., Christie, J., Zare, S., McMurray, J. and Dargie, H. (1999). A comparison of the reproducibility and the sensitivity to change of visual analogue scales, borg scales, and likert scales in normal subjects during submaximal exercise. Chest 116:1208–1217.

Harreby, K. R., Kundu, A., Geng, B., Maciejasz, P., Guiraud, D., Stieglitz, T., Boretius, T., Yoshida, K. and Jensen, W. (2012). Recruitment selectivity of single and pairs of transverse, intrafascicular, multi-channel electrodes (TIME) in the pig median nerve. Annual Conference of the International Functional Electrical Stimulation Society, IFESS, Banff, Alberta, Canada.

Kaczmarek, K. A., Webster, J. G., Bach-y-Rita, P. and Tompkins, W. J. (1991). Electrotactile and vibrotactile displays for sensory substitution systems. IEEE Trans. Biomed. Eng. 38:1–16.

Karl, A., Mühlnickel, W., Kurth, R. and Flor, H. (2004). Neuroelectric source imaging of steady-state movement-related cortical potentials in human upper extremity amputees with and without phantom limb pain. Pain 110:90–102.

Kooijman, C. M., Dijkstra, P. U., Geertzen, J. H. B., Elzinga, A. and Van Der Schans, C. P. (2000). Phantom pain and phantom sensations in upper limb amputees: An epidemiological study. Pain 87:33–41.

Kundu, A., Harreby, K. R. and Jensen, W. (2012). Comparison of median and ulnar nerve morphology of Danish landrace pigs and göttingen mini pigs. Annual Conference of the International Functional Electrical Stimulation Society, IFESS, Banff, Alberta, Canada.

Lewis, J. R. (1995). IBM computer usability satisfaction questionnaires: psychometric evaluation and instructions for use. International Journal of Human-Computer Interaction, 7(1):57–78.

Lotze, M., Flor, H., Grodd, W., Larbig, W. and Birbaumer, N. (2001). Phantom movements and pain an fMRI study in upper limb amputees. Brain 124:2268–2277.

Marks, L. and Gescheider, G. (2002). Psychophysical scaling. In: Stevens' handbook of experimental psychology Psychophysical scaling.

Pfeiffer, E. A. (1968). Electrical stimulation of sensory nerves with skin electrodes for research, diagnosis, communication and behavioral conditioning: A survey. Med & Biol Engng 6:637–651.

Prior, R. E., Lyman, J., Case, P. A. and Scott, C. M. (1976). Supplemental sensory feedback for the VA/NU myoelectric hand. Background and preliminary designs. Bull Prosthet Res 10:170–191.

Ramachandran, V. S., Rogers-Ramachandran, D., Stewart, M. and Pons, T. P. (1992). Perceptual correlates of massive cortical reorganization. Science 258:1159–1160.

Riso, R. R. (1999). Strategies for providing upper extremity amputees with tactile and hand position feedback – moving closer to the bionic arm. Technol Health Care 7:401–409.

Rossini, P. M., Micera, S., Benvenuto, A., Carpaneto, J., Cavallo, G., Citi, L., Cipriani, C., Denaro, L., Denaro, V., Di Pino, G., Ferreri, F., Guglielmelli, E., Hoffmann, K., Raspopovic, S., Rigosa, J., Rossini, L., Tombini, M. and Dario, P. (2010). Double nerve intraneural interface implant on a human amputee for robotic hand control. Clin Neurophysiol 121:777–783.

Shannon, G. F. (1979). A myoelectrically-controlled prosthesis with sensory feedback. Med Biol Eng Comput 17:73–80.

Stevens, S. S. (1957). On the psychophysical law. Psychol Rev 64:153–181.

Stewart, J. D. (2003). Peripheral nerve fascicles: Anatomy and clinical relevance. Muscle and Nerve 28:525–541.

Szeto, A. and Saunders, F. A. (1982). Electrocutaneous stimulation for sensory communication in rehabilitation engineering, IEEE Trans. Biomed. Eng. BME-29:300–308.

Weeks, S. R., Anderson-Barnes, V. C. and Tsao, J. W. (2010). Phantom limb pain: Theories and therapies. Neurologist 16:277–286.

9

A New Treatment for Phantom Limb Pain Based on Restoration of Somatosensory Feedback Through Intraneural Electrical Stimulation

Guiseppe Granata[1,2,*], Winnie Jensen[3], Jean-Louis Divoux[4], David Guiraud[5], Silvestro Micera[6,7], Xavier Navarro[8,9], Thomas Stieglitz[10,11,12], Ken Yoshida[13] and P. M. Rossini[1,2,*]

[1]Fondazione Policlinico Universitario A. Gemelli – IRCCS, Italy
[2]Università Cattolica del Sacro Cuore, Italy
[3]Department of Health Science and Technology, Aalborg University, Denmark
[4]Axonic, MXM, France
[5]LIRMM, University of Montpellier – INRIA, CNRS, Montpellier, France
[6]Bertarelli Foundation Chair in Translational Neuroengineering, Centre for Neuroprosthetics and Institute of Bioengineering, School of Engineering, École Polytechnique Fédérale de Lausanne (EPFL), Lausanne, Switzerland
[7]The BioRobotics Institute, Scuola Superiore Sant'Anna, Pisa, Italy
[8]Institute of Neurosciences, Department of Cell Biology, Physiology and Immunology, Universitat Autònoma de Barcelona, Bellaterra, Spain
[9]Centro de Investigación Biomédica en Red sobre Enfermedades Neurodegenerativas (CIBERNED), Bellaterra, Spain
[10]Laboratory for Biomedical Microsystems, Department of Microsystems Engineering-IMTEK, Albert-Ludwig-University of Freiburg, Freiburg, Germany
[11]BrainLinks-BrainTools, Albert-Ludwig-University of Freiburg, Freiburg, Germany
[12]Bernstein Center Freiburg, Albert-Ludwig-University of Freiburg, Freiburg, Germany

[13]Department of Biomedical Engineering, Indiana University – Purdue University Indianapolis
E-mail: granata.gius@gmail.com
*Corresponding Authors

9.1 Introduction

Phantom limb pain (PLP) is a frequent consequence of amputation and can be defined as pain in a body part that is no longer present. Although phantom limb pain occurs in up to 85% of patients after amputation (Sherman and Sherman, 1983), the characteristics of this pain vary drastically. Throbbing, piercing and needle sensations are among the most commonly used descriptors of PLP. The heterogeneity of phantom limb pain is widely acknowledged, and it is recognized that the disorder likely arises from multiple mechanisms (Flor et al., 2006). PLP is commonly considered as neuropathic and the general view now is that many changes along the neuraxis, both peripheral and central, contribute to the experience of PLP. A gold standard for PLP treatment is still lacking since several studies, including large surveys of patients with amputations, have shown that most treatments are ineffective (Griffin and Tsao, 2014). Available therapies include acupuncture, deep brain stimulation (Flor, 2002), mirror/virtual reality therapies (Chan et al., 2007; Mercier and Sirigu, 2009), mental imagery (MacIver et al., 2008), transcutaneous nerve stimulation (Katz and Melzack, 1991), deep brain, motor cortex, spinal cord, dorsal root ganglion and peripheral nerve stimulation (Krainick et al., 1980; Bittar et al., 2005; Ahmed et al., 2011), drugs such as carbamazepine, opioids (Wu et al., 2008), gabapentin (Bone et al., 2002), amitriptyline (Robinson et al., 2004), calcitonin (Jaeger and Maier, 1992), ketamine (Eichenberger et al., 2008) and memantine (Nikolajsen et al., 2000), local anaesthesia, sympathectomy, dorsal root entry-zone lesions, cordotomy and rhizotomy (see Sherman et al., 1980; Sherman, 1997).

Following limb amputation and, consequently, truncation of nerves at the stump level, a part of the nerve fibers originally directed to the lost part of the body degenerates (because of retrograde Wallerian degeneration); however, a significant part of them survives in the residual portion of the nerve. Thanks to the rapid development of neural interfaces for the peripheral nervous system, e.g. intraneural multichannel electrodes (Navarro et al., 2005), it was recently demonstrated that it is possible to induct meaningful sensations (Dhillon et al., 2005) in the phantom hand of amputees by stimulating the surviving fibers through these interfaces.

Peripheral nerve stimulation was the first clinical application of the gate control theory proposed by Melzack and Wall (1965). This technique requires the implantation of electrodes near, across or along a nerve trunk (see Rasskazoff and Slavin, 2012) to provide stimulation-induced paresthesia. Anecdotical reports showed efficacy of peripheral nerve stimulation in PLP treatment.

Some studies have demonstrated a positive effect of providing sensory feedback from the lost part of the body to alleviate PLP, e.g. through mirror/virtual reality therapies, myoelectric prosthesis or daily discrimination training of surface stimuli, electrical or mechanical, applied to the stump (Antfolk et al., 2010; Antfolk et al., 2012; Börjman et al., 2012).

In the frame of a European founded project (called TIME), we implemented a system able to perform a peripheral nerve stimulation, providing, at the same time, meaningful somatosensory feedback from the lost hand in amputees (Raspopovic et al., 2014). This is a pilot study aiming to test the feasibility, the safety of this system and its efficacy in treating PLP.

9.2 Methods

All procedures were approved by the local Ethics Committee and, being an experimental trial using non-CE marked medical devices, also by the Italian Ministry of Health. An informed consent was signed by the patient before beginning the trial.

The system that we implemented was composed of two different medical devices:

(1) intra-fascicular multichannel electrodes (TIMEs)
(2) multichannel stimulator system (Stim'ND)

The system was tested in a 34-year-old male with a traumatic transradial (proximal third of the forearm) amputation of the left arm in January 2004. About a week after the amputation, the patient began to present a painful phantom limb syndrome, which at the time of the trial was reported as a constant sensation of "very intense painful clenched fist" of the amputated phantom hand, almost constant, worsening with the cold. On a scale from 0 (no pain) to 10 (maximum pain imaginable), the patient reported a constant pain intensity of 9. Apart from the amputation and consequent PLP, the clinical condition of the patient was unremarkable.

The TIMEs consist of a 10 μm thick polyimide-based substrate with seven active electrode sites on each side of a loop-like structure and two reference electrodes, connected through a ceramic adaptor with a 18 mm long polyimide ribbon ending with a 16 pole circular connector (NPC-16, Omnetics Inc., Minneapolis, USA). Four electrodes (Boretius et al., 2010) were surgically implanted into the median and ulnar nerves of the amputee at the arm level. In both nerves, the two TIMEs were implanted, one distally (closer to the elbow) and the other proximally (closer to the axilla).

To implant the TIMEs, in the operating room, a 15 cm long incision was made in the left arm of the patient (along the internal bicipital groove) in order to expose the nerves. The surgical intervention was performed in general anesthesia. The median and ulnar nerves were dissected from the surrounding tissue along all the length of the opening. In order to implant the TIMEs, the same following procedure was performed for each electrode. A short incision was made in the epineurium of selected nerves to have visual feedback on the location of individual fascicles. Then, using a guiding needle (Ethicon needle diameter 125 μm attached to a 10-0 Prolene suture), each polyimide loop was implanted transversally (with respect to the longitudinal direction of the nerve) to place the active sites inside the nerve, inside or between fascicles. In order to check the positioning of the electrodes, inside the operating room, each TIMEs was connected with the Stim'ND stimulator to assess the electrode–tissue impedance and verify that the active sites had been properly placed inside the nerve. After the check, in the case of proper positioning, the electrode was anchored to the epineurium (with 8-0 non-absorbable Silk suture) and surrounding tissues (with 4-0 non-absorbable Silk suture) to avoid migration. A subcutaneous pocket was created to house the lead-out cables of the TIMEs and, after a small skin incision, the electrode was externalized to make it available for percutaneous connection with the stimulator. Finally, the muscle and tissue were repositioned in their original location and the wound was closed with 3-0 non-absorbable Ethilon suture. Each cable was fixed at the exit point (3-0 non-absorbable Silk sutures).

The surgical operation lasted 7 hours. The participant was discharged from the hospital 48 hours later with no post-surgical complications.

The Stim'ND is a multichannel electrical stimulator that, connected with the TIMEs, was able to deliver electrical stimulation to the nerve through the active sites of TIMEs, individually or simultaneously, with an amplitude range of 20 μA–5.1 mA (step size of 20 μA between current levels) and a pulse width range of 1–511 μs (minimum step size of 1 μs).

As previously stated in the introduction, it is possible to elicit sensations in the phantom hand by stimulating the nerve fascicles (with intraneural multichannel electrodes) originally directed to the hand before the amputation. However, it is impossible to know in advance which type of feedback and where it is possible to evoke these sensations. The only possibility to discover the potential of the system is to perform a mapping of each electrode and corresponding contacts (a process called sensation characterization) in order to know the following:

- the location, type and strength of the generated sensation with respect to the active stimulation sites used to generate them;
- the lower (thresholds) and upper (saturation) limits of the current to be delivered in order to induce meaningful sensations (defined, respectively, as the lowest stimulus pulse charge at which the subject reliably feels a sensation and the pulse charge at which the sensation becomes close to uncomfortable or painful);
- which electrode channels are useful and able to induce reliable sensory feelings in the patient's phantom hand (the sensation is considered reliable if, when using the same characteristics of current, the generated sensation is the same considering the strength, the location and the type in at least two out of three repeated trials of stimulation; the charge delivered for the three repeated trials is between the lower thresholds and upper limits).

For this purpose, each channel of each electrode was connected with the electrical stimulator in order to deliver short trains of current of variable intensity and duration. More precisely, monopolar, charge-balanced, biphasic and rectangular stimulation pulses were applied; the single pulse duration varied between 25 and 300 μs (steps of 25 μs) and the pulse amplitude between 40 and 300 μA (steps of 20 μA), and the pulses were delivered with a frequency of 50 Hz over a 500 ms time window. The stimulator – and consequently the parameters of the current to be injected into the nerves – was managed by the operators, thanks to a specific software. The procedure was performed as follows. First, train of current of increasing amplitude from 40 to 300 μA (steps of 20 μA) with a fixed pulse duration (the lowest, i.e. 25 μs) were delivered. After each stimulus, the patient was asked to provide his feedback on the type, strength and location of the sensation (if any) through a computer interface properly designed for the study. If no sensation was reported, a second train of stimuli with the same range of amplitude but with a higher step duration of 25 μs was delivered. The same procedure was repeated progressively increasing the duration of the current (never exceeding the maximum chemical safe charge injection limit per electrode site, i.e. 120

nC) until the patient felt a mild sensation in the phantom hand for the first time. In this case, the last train of stimuli was repeated twice with the same characteristics and, if the same sensation was reported at least once, the pattern of electrical parameters able to generate this sensation was identified as the lowest threshold and coupled with the type, the localization and the intensity of the sensation reported by the patient. The active sites not able to elicit any sensation using the maximum charge possible, i.e. 120 nC, were discarded from further use.

In order to determine the upper limit of sensation, we continued to deliver trains of increasing current until the generated sensation became uncomfortable. Also, in this case, the procedure was repeated twice with the same characteristics and, if the same sensation was reported at least once, the pattern of electrical parameters able to generate the uncomfortable sensation was identified as the upper limit. Since minor changes of the electrode–tissue interface can be expected during the weeks following surgery (i.e. formation of fibrosis around the TIME device and possible micro-motion of the polyimide loop placed inside the nerve), this procedure was repeated once a week to ensure that the stimulation levels were appropriate.

When the lower threshold and upper limits were identified, a level of current (in the range previously defined between the lower threshold and the upper limit) able to evoke sensation perceived as medium-strength (according to a scale from 0 to 10: 0 corresponding to no sensation and 10 to pain) was delivered and the corresponding location, quality and kind of evoked sensation were characterized.

Finally, strategies of simultaneous synchronous multichannel electrical stimulation (i.e. stimulation of median and ulnar nerves with two or more channels in parallel and simultaneously) were explored. As the number of combinations of electrode channels was extremely large and time-consuming, we first proceeded combining two channels of electrodes implanted in the two different nerves (i.e. one in the median and one in the ulnar nerves) choosing channels able to generate stable and reliable sensations with a minimum amount of charge injected (among those with the same kind of sensory feeling generated, i.e. the channel with the lower “lower threshold”). In order to try to cover a wider area with the evoked sensation of the phantom hand, we proceeded adding one or more channels in different combinations with the aim of reaching the widest surface possible according to the feedback of the patient. Since minor changes to the electrode–tissue interface can be expected, this procedure was also repeated once a week.

The TIMEs were explanted 30 days after the implantation for reasons related to ethical and legal authorization.

9.2.1 Therapy

The most efficacious train of multichannel stimulation was used for pain treatment. More precisely, the stimulation pattern delivered to the patient as therapy was represented by sequences of 15 biphasic pulse trains lasting 0.5 seconds. The time interval between two consecutive trains was 1 second, while that between two consecutive sequences was 10 seconds. The therapy consisted of three sessions of 30 minutes (9 minutes stimulation, 1 minute rest) each separated by 5 minutes. Every session lasted 100 minutes. The daily session of treatment was repeated with the selected set of stimulation parameters and active sites that produced reliable, distinct and the most comfortable and pain-releasing sensations while injecting the minimum amount of charge. The stimulation sessions were performed for 10 days.

Pain assessment

To assess the effect of treatment, we used three different questionnaires: the abbreviated version of the McGill Pain Questionnaire (sfMcGill), the present pain intensity scale (PPI) and the pain visual analogue scale (VAS). In addition, the participant's own qualitative descriptions of his pain perception were recorded in an open-ended session. The questionnaires were used the first time before the surgical implantation of TIMEs, second time after implant of the TIME devices (but before starting the treatment) and daily where we applied repeated stimulation sessions (either after the session only or both before and after the session), and on day 30 (the day of the removal of the TIMEs).

9.2.2 Assessment of Cortical Organization

Cortical reorganization was recorded with two different methods: electroencephalography (EEG) and somatosensory evoked potentials (SEP). Both the examinations were performed before and after the therapy.

9.2.2.1 EEG

EEG signals were recorded using a cap with 31 recording electrodes placed according to the positions of the 10–20 international system (excluding Fpz and Oz). The reference was at Fz and the ground at Fpz. An automatic artifact rejection algorithm excluded from the average all runs containing transient exceeding +65 mV at any recording channel. Data were processed in Matlab R2011b (MathWorks, Natick, MA) using scripts based on EEGLAB 11.0.5.4b toolbox (Swartz Center for Computational Neurosciences, La Jolla, CA). EEG signals were bandpass filtered from 0.1 to 47 Hz using a finite impulse response (FIR) filter. Imported data were divided in 2 second duration epochs and visible artifacts in the EEG recordings (i.e. eye movements, cardiac activity and scalp muscle contraction) were removed using an independent component analysis (ICA) procedure (Formaggio et al., 2013; Hoffmann et al., 2008; Jung et al., 2000) performed using the Infomax ICA algorithm (Bell et al., 1995) as implemented in the EEGLAB. Artifact-free EEG signals were used for the further analyses.

The recordings were performed in resting state, with the patient seated in an arm chair for 5 minutes with eyes open and 5 minutes with eyes closed.

EEG power spectral density

We selected artifact-free EEG epochs and calculated the spectral density (EEG signal bandwidth from 0.5 to 45 Hz with a resolution of 0.5 Hz, frequency bands at delta: 2–4 Hz, theta: 4–8 Hz, alpha 1: 8–10.5 Hz, alpha 2: 10.5–13 Hz, beta 1: 13–20 Hz, beta 2: 20–30 Hz and gamma: 30–45 Hz).

EEG cortical source analysis

We estimated the three-dimensional distribution of the EEG activity by the use of the standardized Low Resolution Electromagnetic Tomography Algorithm (sLORETA). The output of the algorithm integrates information on localization of signal sources with the brain tomography. Selected artifact-free EEG segments were used for calculating the sLORETA intracranial spectral density, with a resolution of 0.5 Hz, from 0.5 to 45 Hz. All EEG data epochs were normalized and re-computed into cortical current density time series at 6239 cortical voxels7. The voxels were collapsed at four regions of interest (ROI) in each hemisphere, central (Brodmann areas: 8–11 and 44–47), frontal (Brodmann areas: 1–4 and 6), parietal (Brodmann areas: 5–7, 30, 39–40 and 43) and occipital (Brodmann areas: 17–19) coded according to the Talairach space. The signal at each cortical ROI consisted of the averaged electric neuronal activities of all voxels belonging to that ROI. The sLORETA has shown to provide improved results over other methodologies, such as the LORETA algorithm. The exact, zero-error localization property of the method has been demonstrated, and no localization bias is introduced even in the presence of measurement or biological noise (Pascual-Marqui et al., 2002;

Pascual-Marqui et al., 2011). Also, localization agreement has been shown with functional magnetic resonance imaging (fMRI), structural MRI, positron emission tomography (PET) and intracranial recordings in humans.

Functional connectivity analysis

Brain connectivity was computed by sLORETA/eLORETA software on 84 ROIs defined according to the 42 Brodmann areas (BAs: 1–11, 13, 17–25, 27–47) for each hemisphere. Among the eLORETA current density time series of the 84 ROIs, intracortical lagged linear coherence, extracted by "all nearest voxels" method, was computed between all possible pairs of the 84 ROIs for each of the seven EEG frequency bands (delta (2–4 Hz), theta (4–8 Hz), alpha 1 (8–10.5 Hz), alpha 2 (10.5–13 Hz), beta 1 (13–20 Hz), beta 2 (20–30 Hz) and gamma (30–45 Hz). The values of connectivity computing between all pairs of ROIs for each frequency band were used as measure of weight of the graph in the follow graph analyses. The lagged linear coherence method was originally developed as a measure of physiological connectivity that was not affected by volume conduction and low spatial resolution. It has been shown to provide an improved connectivity measure8 in comparison to the imaginary coherence method (Nolte et al., 2004).

Graph analysis

A network is a mathematical representation of a real-world complex system and is defined by a collection of nodes (vertices) and links (edges) between pairs of nodes. Nodes in large-scale brain networks usually represent brain regions, while links represent anatomical, functional or effective connections (Friston et al., 1994), depending on the dataset. Anatomical connections typically correspond to white matter tracts between pairs of grey matter (cortical areas or subcortical relays) brain regions.

Functional connections correspond to magnitudes of temporal correlations in activity and may occur between pairs of anatomically unconnected regions. The nature of nodes and links in brain networks is determined by combining brain mapping methods, anatomical parcellation schemes and measures of connectivity. Nodes should ideally represent brain regions with coherent patterns of extrinsic anatomical or functional connections (Rubinov et al., 2006; Sporns et al., 2006). Two core measures of graph theory were computed: characteristic path length (L) and weighted clustering coefficient, representative of global connectedness and local interconnectedness, respectively (Watts et al., 1998). The length of a path is indicated by the number of connections it contains. The characteristic path length L (averaged shortest path length between all node pairs) is an emergent property of the graph, which indicates how well its elements are integrated/interconnected. The mean clustering coefficient is computed for all nodes of the graph and then averaged. It is a measure for the tendency of network elements to form local clusters 14. The respective distributions of global (L-random) and local (C-random) connectedness values were calculated and averaged to obtain a mean value for each core measure. The scale-free value, Gamma (γ) was evaluated from the normalized C (C/C-random) and Lambda (λ) from the normalized L (L/L-random). "Small-worldness" (Sigma σ) is the γ:λ ratio and is used to describe the balance between the local connectedness and the global integration of a network. When this ratio is larger than 1, a network is said to have small-world properties.

9.2.2.2 SEP

To assess the cortical topography and responsiveness to sensory stimulation, we analyzed somatosensory evoked maps. To obtain SEP, we stimulated the median nerve at the elbow. The SEP amplitudes were measured in relation to baseline recordings. SEP cortical components were identified on the basis of their latency and polarity, and we created cortical maps based on the N20 component. The N20 component corresponds to arrival of the peripheral nerve stimulation at the primary somatosensory area.

Somatosensory evoked potentials were performed by using a Micromed System Plus Evolution, Mogliano Veneto, Italy. The somatosensory evoked potentials were recorded using a cuff with 31 recording electrodes placed according to the positions of the 10–20 international system 31 electrodes, Fp1, Fp2, AF3, AF4, F7, F3, Fz, F4, F8, FC5, FC1, FC2, FC6, T3, C3, C7, C4, T4, CP5, CP1, CP2, CP6, T5, P3, P7, P4, T6, PO3, PO4, O1 and O2. The common reference was placed at the vertex and the ground at the forehead.

The signal was amplified and filtered (bandpass 0.3–70 Hz). An automatic artifact rejection algorithm was used in order to exclude all runs containing transient exceeding +65 mV at any recording channel from the average. Peripheral nerve stimulation was performed with a surface bipolar electrode stimulating the median nerve at the elbow, medially to the distal tendon of biceps brachii. The electrical stimulation was performed with the stimulator integrated in the Micromed System, and a pulse intensity able to generate a clear motor twitch of flexors carpi and/or finger muscles was used. The frequency of the stimulation was 1 Hz. Two separate averages – each composed by 500 stimuli – were performed. If needed, according to the operator's judgment on the quality of the somatosensory evoked potentials recorded, one or more averages were added. The whole procedure was repeated on the amputee side before and after the therapy and on the spared side only before the therapy.

9.3 Results

The surgical implantation and removal of TIME electrodes on an amputee was smooth and efficient after its careful planning. No complications related to the surgical procedure and daily stimulation were observed. In particular, no complications related to nerve injury, infection, bleeding and relocation of the electrodes were observed.

The four implanted electrodes functioned during all the trials. It was possible to stimulate the four implanted nerves using all channels except seven contacts. The intraneural stimulation was able to evoke different sensations referred by the patients as "touch", "pressure", "vibration", "a touching wave" and "warm". The more common reported sensations were "touch" and "vibration". The sensations were primarily located on the palm, thumb, index and little finger of the phantom hand. The anatomical distribution of this sensation was always congruent with the electrode used: electrodes inserted on median nerve generated sensation in the median nerve innervated part of the phantom hand and electrodes implanted on ulnar nerve evoked sensation on the ulnar innervated portion of the phantom hand. Some channels, when stimulated, produced an evident muscular twitch of the stump. The combined stimulation of two or more different channels usually had two common effects:

(1) The area of the phantom hand where the patient felt the sensation was not simply the sum of the two portions of hand stimulated by using only one of the channels.
(2) The current threshold needed to evoke a sensation was usually lower than the threshold of the same electrode when used alone.

The charge required to elicit a specific sensation increased over time; however, the quality and strength of the sensation generated remained stable over time.

Many combinations of channels were tried and a combination of five different active sites was selected for the treatment. This combination of channels produced a reliable, distinct, pain releasing and comfortable sensation perceived by the patient as "a touching wave" covering the majority of the palmar side of his phantom hand (Figure 9.1).

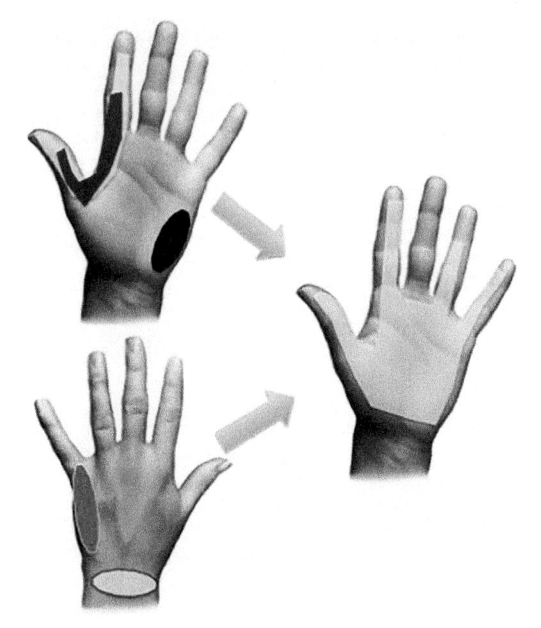

Figure 9.1 Schematic representation of the areas of the patient's phantom hand involved by the sensation during intraneural stimulation. On the left side of the picture, the areas of the phantom hand where the patient felt the sensation stimulating with five channels individually can be seen. On the right side of the picture, the area of phantom hand where the patient felt the sensation by stimulating simultaneously with the same five channels can be seen.

During the 10 days of treatment, the patient described a significant and progressive decrease of his phantom limb pain (Figure 9.2) that was referred as less "stabbing", "sickening", "sharp", "fearful", "gnawing", "cramping", "hot", "aching" and "heavy". The patient experienced a maximum decrease of pain intensity from 9 to 4/5 according to the VAS. A sensation of partial opening of the phantom "clenched fist" was also referred by the patient together with the pain decrease. However, the effect of treatment was of short duration lasting only some hours after therapy; during the first three days of therapy (lowest VAS score between 6 and 7, PPI score 4 and McGill score between 12 and 20), the pain came back to normal before the patient went to sleep, whereas during the last days of therapy, the relief from pain lasted until the patient went to sleep and pain came back to normal when the patient woke up (lowest VAS score between 4 and 5, PPI score 3 and McGill scores between 6 and 9).

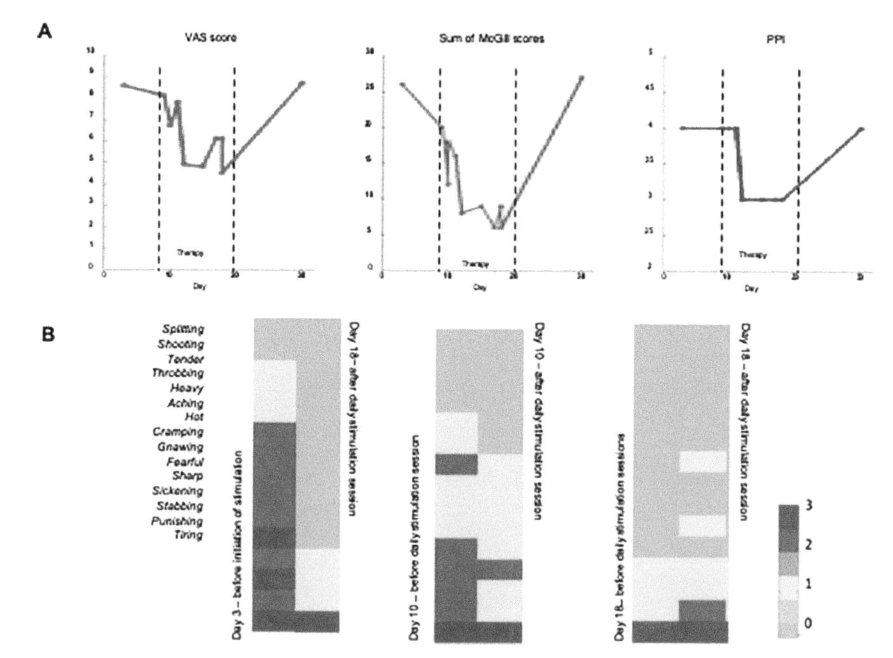

Figure 9.2 Results of questionnaires for PLP evaluation. (A) A clear reduction of PLP is evident in all three questionnaires: VAS, McGill and PPI. (B) Results of McGill more in detail: a clear reduction of different qualities of pain during the treatment is evident.

The TIMEs were removed 30 days after the implantation without any side effects.

From a neurophysiological point of view, the somatosensory evoked potentials mapping showed a modification of the cortical topography in the central–parietal areas contralateral to the amputation site, i.e. evaluated while stimulating the left median nerve (Figure 9.3A). For comparison, we included a map on the evoked potentials while stimulating the right median nerve. The EEG source analysis (Figure 9.3B–C) showed a widespread reduction of delta activity and a significant increase of alpha activity in the central–parietal areas in both hemispheres following the repeated stimulation sessions, indicating a shift of the EEG activity towards normal states.

Finally, the analysis of the cortical connectivity showed a less random architecture of cerebral networks following the repeated stimulation sessions, evident as a reduction on delta band and an increase on alpha band frequencies (Figure 9.3D).

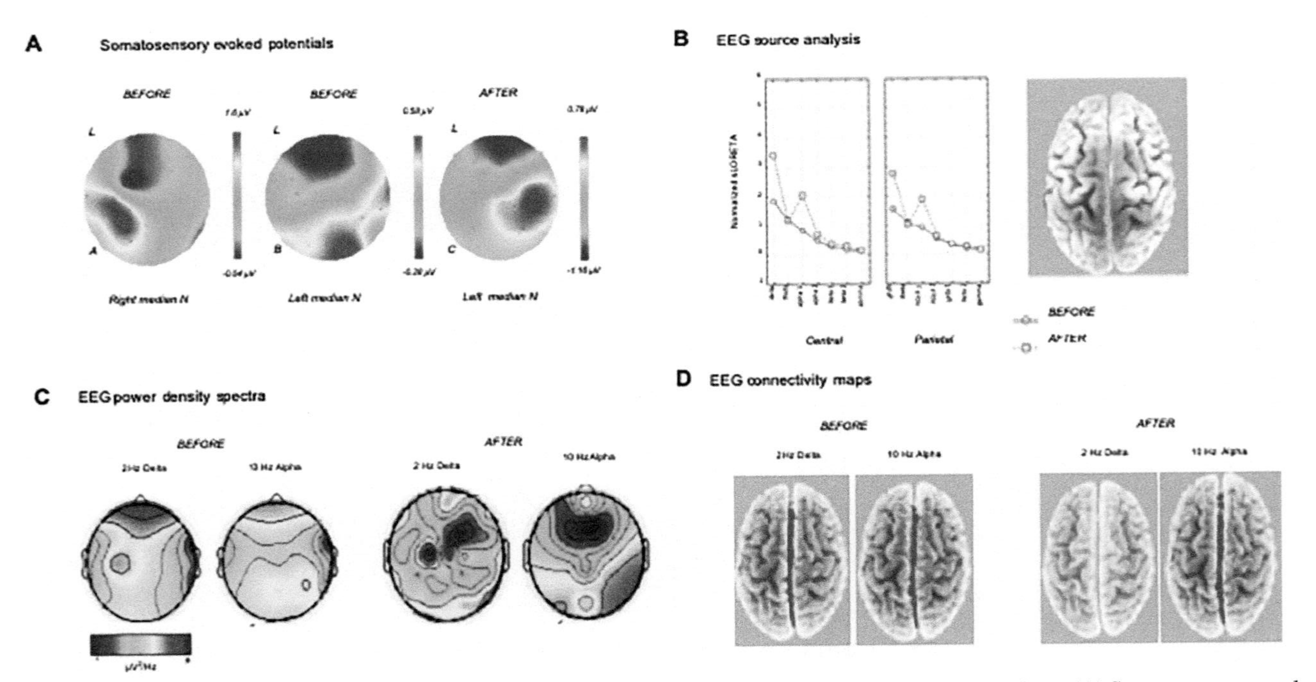

Figure 9.3 Assessment of cortical organization and reorganization before and after repeated stimulation sessions. (A) Somatosensory evoked cortical potentials evaluated before and after the repeated stimulation sessions. For comparison, we included a map on the evoked potentials while stimulating the right median nerve. (B) EEG current sources determined before and after the repeated stimulation sessions. The analyses in A and B show a modification of the cortical topography in the central–parietal areas contralateral to the amputation. (C) Analysis of the EEG power and (D) cortical connectivity before and after the repeated stimulation sessions. The EEG analysis in C and D demonstrated a scattered reduction of delta activity and increase of alpha activity, indicating a shift of the EEG activity towards normal states and towards less random architecture.

9.4 Discussion

To the best of our knowledge, in this study, we present for the first time the results of a new approach, based on peripheral nerve stimulation, for PLP treatment. Our system combines the characteristics of peripheral nerve stimulation with the new kind of intraneural interface that makes it possible to evoke somatosensory sensation on the phantom hand of amputees. The philosophy at the base of this new system was the possibility to combine two kinds of treatment that individually already demonstrated, although in a very small number of cases, efficacy in dealing with intractable (or poorly responsive) PLP. Paresthesias are assumed essential for pain relief when using conventional spinal cord, dorsal root ganglion or peripheral nerve stimulation systems (Kumar et al., 1998; North et al., 2006). The other assumption of these kinds of systems is the possibility to cover the areas of the body involved by pain with paresthesia generated by electrical stimulation. In the design of the protocol, along the lines of already existing and previously cited kinds of stimulations, we decided to choose a pattern of stimulation able to cover, as much as possible, the phantom hand surface, i.e. the painful area in our amputee patient. In this perspective, median and ulnar nerves were chosen for the implantation of TIMEs because their innervation territory covers almost entirely the palmar and the finger sensory fields. In addition to the previous anatomical consideration, a practical aspect was also taken into account in choosing nerves to be implanted; given the level of amputation with a single incision at the medial aspect of the arm, it was possible to expose both median and ulnar nerves. To implant the radial nerve would have meant a second incision and an even greater duration of surgery.

Paresthesias can be uncomfortable for some patients and sometimes can limit the acceptable amplitude of stimulation during conventional spinal cord, dorsal root ganglion or peripheral nerve stimulation systems (Kuechmann et al., 2009). One of the advantages of our system is that we are able to avoid this unpleasant sensation by substituting it with a natural, meaningful and pleasant sensation in the missing part of the body. Moreover, the selective stimulation of sensory fibers bypass the difficulty to stimulate mixed sensory and motor nerve, avoiding recruitment of motor fibers. The latter is one of the reasons why the electrical stimulation of the end "fields" of peripheral nerves technique was developed and, also if in its infancy, is growing as an effective option for local generation of paresthesia. However, in amputees, this is not an option due to the absence of the area involved from pain. All conventional electrical stimulation treatments (spinal cord, peripheral and dorsal root ganglion) require implantation of a dedicated device with the only purpose of pain treatment. The future of prosthetics is going toward the use of new generation prosthesis that can be controlled bidirectionally by the users. This kind of prosthesis is based on the possibility to provide a somatosensory feedback to the amputees, and that possibility is based on the implant of the same (or similar) electrode used in our system (Raspopovic et al., 2014; Tan et al., 2014). So, in the future perspectives, the great advantage of our approach is that the same system will be used both for PLP treatment and for the functioning of this new generation prosthesis.

During the 10 days of treatment, the patient experienced a decrease in his phantom pain intensity together with a change in the qualitative perception of his pain. However, this was only a transient effect (only for few hours following the treatment). It is outside the purpose and possibilities of this work to understand the mechanism of pain decrease and at what level of the nervous system the stimulation might have acted, but some considerations can be made. Peripheral nerve stimulation represented the first clinical application of the gate control theory proposed by Melzack and Wall (1965). According to this theory, the A-beta fibers interact within the substantia gelatinosa of the posterior horn in the spinal cord with inhibitory interneurons able to influence the wide dynamic range neuron, where both the large and small pain fibers synapse, closing the gate and inhibiting the conduction of pain (Treede, 2016). Although some recent evidence has questioned this theory, the concept of early large fiber recruitment inhibiting small fiber conduction remains the basis of the theory of electrical stimulation for pain treatment, and recent studies demonstrated that peripheral nerve stimulation is able to suppress subjective complaints of pain associated with noxious laser-induced nociception (Ellrich and Lamp, 2005). The kind of sensation mainly evoked with intraneural stimulation in our study was touch and vibration, and for this reason, it is possible to hypothesize that with our system, large diameter sensory fibers, like A-beta fibers, were the most likely recruited (probably because of their lower excitability threshold). According to this assumption, it is possible to speculate that one of the possible effects of our system was mediated by the activation of these fibers with a mechanism similar to what usually happens with peripheral nerve or other kinds of electrical stimulations (e.g. spinal cord or dorsal root ganglion stimulations). The transient duration of pain relief seems in accordance with this mechanism. However, other mechanisms and levels of stimulus-induced plastic changes, both peripheral and central, cannot be excluded. In particular, our results showed plastic changes in the primary somatosensory area and a general increase of alpha band power density and connectivity following the trial that could have contributed to the pain relief.

Our study confirmed the possibility, also after years from the amputation, to deliver intraneural electrical stimulation and to generate meaningful and comfortable sensations perceived from the phantom hand by amputees.

The majority of the channels of TIMEs remained functional during the whole trial and the system was safe. No side effects related to the implant/explant of TIMEs and repeated intraneural stimulation were observed within a three-year follow-up.

Since the implant of the TIMEs cannot be fully controlled to target specific sensory nerve fibers, we had to perform a mapping procedure in order to match the different channels of the four electrodes with the corresponding potentially evoked sensations. This is a highly time-consuming procedure but, without a conceptual redesign of described procedures, should be considered a necessary pre-condition for the proposed treatment.

The charge of electrical stimulation increased over time, likely due to the ongoing healing process of the tissue and formation of fibrosis. For this reason, a quick check of threshold at the beginning of each week was needed. This is an issue that should be taken into account for future clinical applications. The chronic application of this kind of treatment remains a big unsolved issue and longer studies should be performed in order to test the stability of the system over time.

Other limitations of the study are the following. (1) Data were obtained only from one patient and thus generalization to a population should be cautioned. However, this was a pilot study, and it is important to underline the strong effort, in terms of time and money, performed in order to obtain this proof of concept. (2) Intensity and dosage of treatment were fixed, and it was not possible, also for ethical reasons, to blind the intervention neither for the participant nor for the experimenters. Moreover, it was not possible to conduct a SHAM stimulation. However, regarding this point, it is important to underline that the simple procedure of intraneural stimulation did not provoke any modification to PLP in our patient.

This study provides the demonstration of feasibility, safety and efficacy of a newly developed system for PLP treatment. Other studies are needed in order to test this treatment in a larger number of patients. Using the same system, different paradigms of stimulation, in terms of frequency and duration of treatment, could also be tested. The ongoing process of development of new generation prosthesis, able to provide a somatosensory feedback, should be considered as an opportunity to test this new therapeutic approach.

References

Ahmed, M. A., Mohamed, S. A. and Sayed, D. (2011). Long-term antalgic effects of repetitive transcranial magnetic stimulation of motor cortex and serum beta-endorphin in patients with phantom pain. Neurol Res 33:953–8.

Antfolk, C., Balkenius, C., Rosén, B., Lundborg, G. and Sebelius, F. (2010). SmartHand tactile display: a new concept for providing sensory feedback in hand prostheses. Scand J Plast Reconstr Surg Hand Surg 44:50–3.

Antfolk, C., Cipriani, C., Carrozza, M. C., Balkenius, C., Björkman, A., Lundborg, G., Rosén, B. and Sebelius, F. (2012). Transfer of tactile input from an artificial hand to the forearm: experiments in amputees and able-bodied volunteers. Disabil Rehabil Assist Technol 8:249–54.

Bell, A. J. and Sejnowski, T. J. (1995). An information-maximization approach to blind separation and blind deconvolution. Neural Comput 7(6):1129–1159.

Bittar, R. G., Otero, S., Carter, H. and Aziz, T. Z. (2005). Deep brain stimulation for phantom limb pain. J Clin Neurosci 12:399–404.

Björkman, B., Rosén, B., Weibull, A., Ehrsson, H. H. and Björkman-Burtscher, I. (2012). Phantom digit somatopy – a functional magnetic imaging study in forearm amputees. European Journal of Neuroscience 36:2098–106.

Bone, M., Critchley, P. and Buggy, D. J. (2002). Gabapentin in postamputation phantom limb pain: A randomized, double-blind, placebo-controlled, cross-over study. Reg Anesth Pain Med 27:481–6.

Boretius, T., Badia, J., Pascual-Font, A., Schuettler, M., Navarro, X. and Yoshida, K. et al. (2010). A transverse intrafascicular multichannel electrode (TIME) to interface with the peripheral nerve. Biosens Bioelectron 26:62–69.

Chan, B. L., Witt, R., Charrow, A. P., Magee, A., Howard, R., Pasquina, P. F., Heilman, K. M. and Tsao, J. W. (2007). Mirror therapy for phantom limb pain. N Engl J Med 357:2206–7.

Dhillon, G. S., Krüger, T. B., Sandhu, J. S. and Horch, K. W. (2005). Effects of short-term training on sensory and motor function in severed nerves of long-term human amputees. J Neurophysiol 93:2625–2633.

Eichenberger, U., Neff, F., Sveticic, G., Bjorgo, S., Petersen-Felix, S., Arendt-Nielsen, L. and Curatolo, M. (2008). Chronic phantom limb pain: The effects of calcitonin, ketamine, and their combination on pain and sensory thresholds. Anesth Analg 106:1265–73.

Ellrich, J. and Lamp, S. (2005). Peripheral nerve stimulation inhibits nociceptive processing: an electrophysiological study in healthy volunteers. Neuromodulation 8:225–32.

Flor, H., Nikolajsen, L. and Staehelin Jensen, T. (2006). Phantom limb pain: A case of maladaptive CNS plasticity? Nat Rev Neurosci 7:873–81.

Flor, H. (2002). Phantom-limb pain: Characteristics, causes, and treatment. Lancet Neurol 1:182–9.

Formaggio, E., Storti, S. F. and Boscolo Galazzo, I. et al. (2013). Modulation of event-related desynchronization in robot-assisted hand performance: Brain oscillatory changes in active, passive and imagined movements. J Neuroeng Rehabil 10:24-0003-10-24.

Friston, K. J. (1994). Functional and effective connectivity in neuroimaging: A synthesis. Hum Brain Mapp 2:55–78.

Griffin, S. C. and Tsao, J. W. (2014). A mechanism-based classification of phantom limb pain. Pain 155:2236–42.

Hoffmann, S. and Falkenstein, M. (2008). The correction of eye blink artefacts in the EEG: A comparison of two prominent methods. PLoS One 3(8):e3004.

Jaeger, H. and Maier, C. (1992). Calcitonin in phantom limb pain: A double-blind study. PAIN 48:21–7.

Jung, T. P., Makeig, S., Humphries, C. et al. (2000). Removing electroencephalographic artifacts by blind source separation. Psychophysiology 37(2):163–178.

Katz, J. and Melzack, R. (1991). Auricular transcutaneous electrical nerve stimulation (TENS) reduces phantom limb pain. J Pain Symptom Manage 6:73–83.

Krainick, J. U., Thoden, U. and Riechert, T. (1980). Pain reduction in amputees by long-term spinal cord stimulation. Long-term follow-up study over 5 years. J Neurosurg 52:346–50.

Kuechmann, C., Valine, T. and Wolfe, D. (2009). Could automatic position-adaptive stimulation be useful in spinal cord stimulation? Eur J Pain 13:S243.

Kumar, K., Toth, C., Nath, R. K. and Laing, P. (1998). Epidural spinal cord stimulation for treatment of chronic pain – some predictors of success. A 15-year experience. Surg Neurol 50:110–120.

MacIver, K., Lloyd, D. M., Kelly, S., Roberts, N. and Nurmikko, T. (2008). Phantom limb pain, cortical reorganization and the therapeutic effect of mental imagery. Brain 131:2181–91.

Melzack, R. and Wall, P. D. (1965). Pain mechanisms: a new theory. Science 150:971-9-79.

Mercier, C. and Sirigu, A. (2009). Training with virtual visual feedback to alleviate phantom limb pain. Neurorehabil Neural Repair 23:587–94.

Navarro, X., Krueger, T., Lago, N., Micera, S., Stieglitz, T. and Dario, P. (2005). A critical review of interfaces with the peripheral nervous system for the control of neuroprostheses and hybrid bionic systems. J Peripher Nerv System 10:229–258.

Nikolajsen, L., Gottrup, H., Kristensen, A. G. and Jensen, T. S. (2000). Memantine (a N-methyl-D- aspartate receptor antagonist) in the treatment of neuropathic pain after amputation or surgery: A randomized, double-blinded, cross-over study. Anesth Analg 91:960–6.

Nolte, G., Bai, O., Wheaton, L., Mari, Z., Vorbach, S. and Hallett, M. (2004). Identifying true brain interaction from EEG data using the imaginary part of coherency. Clin Neurophysiol 115(10):2292–2307.

North, R., Kidd, D., Olin, J., Sieracki, F. and Petrucci, L. (2006). Spinal cord stimulation for axial low back pain: a prospective controlled trial comparing 16-contact insulated electrodes with 4-contact percutaneous electrodes. Neuromodulation 9:56–67.

Pascual-Marqui, R. D., Lehmann, D. and Koukkou, M. et al. (1952). Assessing interactions in the brain with exact low-resolution electromagnetic tomography. Philos Trans A Math Phys Eng Sci 369:3768–3784.

Pascual-Marqui, R. D. (2002). Standardized low-resolution brain electromagnetic tomography (sLORETA): Technical details. Methods Find Exp Clin Pharmacol 24 Suppl D:5–12.

Raspopovic, S., Capogrosso, M., Petrini, F. M., Bonizzato, M., Rigosa, J., Di Pino, G., Carpaneto, J., Controzzi, M., Boretius, T., Fernandez, E., Granata, G., Oddo, C. M., Citi, L., Ciancio, A. L., Cipriani, C., Carrozza, M. C., Jensen, W., Guglielmelli, E., Stieglitz, T., Rossini, P. M. and Micera, S. (2014). Restoring natural sensory feedback in real-time bidirectional hand prostheses. Sci Transl Med 6:222ra19.

Rasskazoff, S. Y., Slavin, K. V. (2012). An update on peripheral nerve stimulation. J Neurosurg Sci 56:279–85.

Robinson, L. R., Czerniecki, J. M., Ehde, D. M., Edwards, W. T., Judish, D. A., Goldberg, M. L., Campbell, K. M., Smith, D. G. and Jensen, M. P.

(2004). Trial of amitriptyline for relief of pain in amputees: Results of a randomized controlled study. Arch Phys Med Rehabil 85:1–6.

Rubinov, M. and Sporns, O. (2010). Complex network measures of brain connectivity: Uses and interpretations. Neuroimage 52(3): 1059–1069.

Sherman, R. A., Sherman, C. J. and Gall, N. G. (1980). A survey of current phantom limb pain treatment in the United States. Pain 8:85–99.

Sherman, R. A. and Sherman, C. J. (1983). Prevalence and characteristics of chronic phantom limb pain among American veterans. Results of a trial survey. Am J Phys Med 62:227–38.

Sherman, R. A. (1997). Phantom limb pain. New York: Plenum, 1997.

Sporns, O. and Honey, C. J. (2006). Small worlds inside big brains. Proc Natl Acad Sci USA 103(51):19219–19220.

Tan, D. W., Schiefer, M. A., Keith, M. W., Anderson, J. R., Tyler, J. and Tyler, D. J. (2014). A neural interface provides long-term stable natural touch perception. Sci Transl Med 6:257ra138.

Treede, R. D. (2016). Gain Control Mechanisms in the Nociceptive System. Pain [Epub ahead of print].

Watts, D. J. and Strogatz, S. H. (1998). Collective dynamics of 'small-world' networks. Nature 393(6684):440–442.

Wu, C. L., Agarwal, S., Tella, P. K., Klick, B., Clark, M. R. Haythornthwaite, J. A., Max, M. B. and Raja, S. N. (2008). Morphine versus mexiletine for treatment of postamputation pain: A randomized, placebo-controlled, crossover trial. Anesthesiology 109:289–96.

10

Future Applications of the TIME

Thomas Stieglitz[1,2,3]

[1]Laboratory for Biomedical Microsystems, Department of Microsystems Engineering-IMTEK, Albert-Ludwig-University of Freiburg, Freiburg, Germany
[2]BrainLinks-BrainTools, Albert-Ludwig-University of Freiburg, Freiburg, Germany
[3]Bernstein Center Freiburg, Albert-Ludwig-University of Freiburg, Freiburg, Germany
E-mail: stieglitz@imtek.uni-freiburg.de

The development of TIME electrodes was driven by the goal to investigate the treatment of phantom limb pain in upper limb amputees by selective intraneural stimulation. Can the experience with the approach be transferred in other applications? What lessons have been learned with the approach of intraneural thin-film-based stimulation electrodes embedded in polymer substrates with respect to other implantation sites, stimulation paradigms, and even other designs?

This chapter will discuss the most obvious next applications and speculate about transfer in other treatment and rehabilitation scenarios in which flexible but mechanically robust nerve interfaces could be advantageous.

Electrical stimulation to deliver sensory feedback (Pasluosta, 2017) is not only limited to patients with limb loss after amputation. Evoked somatosensory potentials (Granata, 2018) of the phantom hand cannot only be used to diminish phantom limb pain (Stieglitz, 2012) but also to deliver signals for sensory feedback in the control of artificial hand prostheses (Micera, 2016). While the benefit of information about grasp force and object compliance during grasping is beyond discussion and was achieved in a subchronic pilot study (Raspopovic, 2014), improvement of proprioception (D'Anna) and embodiment (Rognini, 2018) based on intraneural sensory stimulation could be shown in chronic implantations in three patients. Since delivery of sensory feedback by intraneural stimulation was beneficial after amputation of the upper limb, the research question arose whether sensory feedback after lower limb amputation would improve the use of leg prostheses. Would there be any additional information beyond the one that leg prostheses users get by mechanical interaction of the stump with the prosthesis socket? Preliminary results from studies over up to 4 months in three patients indicate that sensory feedback delivered via four TIME in the residual nerves in the stump does not only reduce energy consume during walking and decreases phantom limb pain (Petrini, 2017 and 2019) but also restores the dexterity, confidence, and ownership in lower-limb amputees (Raspopovic, 2017).

Further transfer scenarios have not been studied yet but include all approaches in which nerves allow transversal implantation by pulling from anatomical access and structure. Linear arrangements or arrays might cover the cross-sectional area of any medium-sized mono- or multifascicular nerve or selectively address subset of fibers in muscles for recording (Farina, 2008) or stimulation. These scenarios include both, control and drive of artificial limb prostheses as well as neuroprothetic control of paralyzed limbs. While early suggestions of double-sided electrode arrays where pure technological design studies (Stieglitz, 2001) on the nerve interface, current technology, and manufacturing readiness levels have matured and allowed chronic first-in-human studies. The requirement of electrode sites on both sides of the implanted probe can be achieved by folding of the thin substrate instead of using complex processes that drive manufacturing costs and quite often result in low yield (Stieglitz, 2002; Poppendieck, 2014). However, for clinical applications wireless systems without the need of percutaneous wires have to be developed and delivered. If the electrode array with many channels and the hermetic package housing electronic circuitry are rather concentrated like in cochlear implants (Lenarz, 2018), direct connection of the array and the hermetic package can be envisioned given that reliable pad-to-pad insulation is delivered over the life-time of the implant (Khan, 2018).

Experiences made with clinical implants that experience mechanical forces due to limb or muscle movements like cardiac pacemakers and deep brain stimulators indicate the necessity of solvable connections to minimize the invasiveness of the surgical procedure and allow for replacement of implant components in the case of (wire) failures. The lack of implantable connectors with about 16 channels or more is currently challenging research groups as well as companies to develop fully implantable systems that would be able to drive several TIMEs at a time with a single implantable pulse generator. Stimulator, package, and feedthrough technology is available (Kohler, 2017) desperately awaiting innovation in the connector development. Challenges in the course of increasing the technology readiness level and commercialization address mainly aspects of longevity. Thin-film electrodes embedded into polyimide substrates have been working reliably over up to 5 months under electrical stimulation in human studies and for more than 2 years in recording studies in nonhuman primates (unpublished data) but they still have to prove their stability. The transition between the micromachined substrate and "real" cables is one of the critical transitions known from many medical and nonmedical developments. Scalability with respect to cables and connectors is very specific for implants since approaches from telecommunication and automotive lack solutions for insulation in humid and salty environment or size restrictions. A possible solution lies in the integration of electronics with multiplexing capabilities and wireless data transmission and energy supply close to the electrode array. Technology is available and solutions could be developed with state of the art knowledge. This work, however, needs more time and money that can regularly be allocated in research project based on public funding. Private-public-partnerships, for example, made in the framework of either strategic or venture capital-based start-up companies, look for prefer to look for application with large patient numbers and high therapeutic need to be able to transfer research ideas into a "mass market" product and have a realistic chance for break even. Different levels of translational research can be foreseen in the preclinical research market or in clinical studies and application. The lowest hanging fruits include the development of flexible probes for the neuroscientific research community that mainly addresses cortical or subcortical structures in rodent animal models. Using insertion tools, polyimide-based probes show foreign body reactions that are comparable to established microprobes or even below (Boehler, 2017). When hearing research goes beyond cochlear implants toward the acoustic nerve (Lenarz, 2018), the TIME approach delivers a good combination of robustness in handling and flexibility at the material-tissue interface. It has all prerequisites to deliver spatiotemporal stimulation patterns in a resolution adequate to restore hearing at a better quality than currently existing auditory brainstem implants. Huge patient numbers would benefit from solutions in the field of bioelectronics medicine (Bouton, 2017). Widespread diseases in aging societies include diabetes, hypertension, rheumatic arthritis, and asthma with patient numbers far above those of neuronal disorders. The vision to interface target nerves close to the end organs or interfere with the autonomous nervous system via the vagus nerve opens many scenarios in which the TIME concept can deliver highly selective nerve interfaces with relatively low degree of invasiveness. However, proof-of-concept studies in small animal models have to validate the general concepts in this relatively new field or research until translational studies can bring hope and new treatment options to these patients.

References

Böhler, C., Kleber, C., Martini, N., Xie, Y., Dryg, I., Stieglitz, T., Hofmann, U. G. and Asplund, M. (2017). Actively controlled release of Dexamethasone from neural microelectrodes in a chronic in vivo study, Biomaterials 129: 176–187.

Bouton, C. (2017). Cracking the neural code, treating paralysis and the future of bioelectronic medicine. J Intern Med 282(1): 37–45.

D'Anna, E., Valle, G., Strauss, I., Patton, J., Petrini, F. M., Raspopovic, S., Granata, G., DiOrio, R., Stieglitz, T., Rossini, P. M. and Micera, S. (2017). Simultaneous tactile and proprioceptive feedback in myo-controlled hand prostheses using intraneural electrical stimulation. Poster presentation at the 47th annual meeting of the Society for Neuroscience (SfN 2017), Washington, D.C., USA, Abstract # 769.03.

Farina, D., Yoshida, K., Stieglitz, T. and Koch, K. P. (2008). Multi-channel thin-film electrode for intramuscular electromyographic recordings, J Appl Physiol 104(3): 821–827.

Granata, G., Di Iorio, R., Romanello, R., Iodice, F., Raspopovic, S., Petrini, F., Strauss, I., Stieglitz, T., Cvancara, P., Giraud, D., Divoux, J.-L., Wauters, L., Hisarray, A., Jensen, W., Micera, S. and Rossini, P. M. (2018). Phantom somatosensory evoked potentials following selective fascicular intraneural electrical stimulation in two amputees. Clin Neurophysiol 129: 1117–1120.

Khan, S., Ordonez, J. S. and Stieglitz, T. (2018). Reliability of spring interconnects for high channel-count polyimide electrodes array. J Micromach Microeng 28 055007 (9 p.).

Kohler, F., Gkogkidis, C. A., Bentler, C., Wang, X., Gierthmuehlen, M., Fischer, J., Stolle, C., Reindl, L. M., Rickert, J., Stieglitz, T., Ball, T. and Schuettler, M. (2017). Closed-loop interaction with the cerebral cortex: a review of wireless implant technology. Brain-Computer Interfaces 4(3): 146–154.

Lenarz, T. (2018). Cochlear implant – state of the art. GMS Curr Top Otorhinolaryngol Head Neck Surg 16:Doc04.

Micera, S., Raspopovic, S., Petrini, F., Carpaneto, J., Oddo, C., Badia, J., Stieglitz, T., Navarro, X., Rossini, P. M. and Granata G. (2016). On the use of intraneural transversal electrodes to develop bidirectional bionic limbs. In: Jaime Ibáñez, José González-Vargas, José María Azorín, Metin Akay, José Luis Pons (eds.), Converging Clinical and Engineering Research on Neurorehabilitation II. Proceedings of the 3rd International Conference on NeuroRehabilitation (ICNR2016), Segovia, Spain. Cham (Switzerland): Springer International Publishing: 737–741.

Pasluosta, C., Kiele, P. and Stieglitz, T. (2017). Paradigms for restoration of somatosensory feedback via stimulation of the peripheral nervous system. Clinical Neurophysiol. DOI: 10.1016/j.clinph.2017.12.027.

Petrini, F. M., Valle, G., Berberi, F., Bortolotti, D., Cvancara, P., Hiairrassary, A., Guiraud, D., Divoux, J.-L., Lesic, A., Stieglitz, T., Micera, S., Raspopovic, S. and Bumbasirevic, M. (2017). Sensory feedback driven by intraneural stimulation allows amputees to reduce energy consume during walking and decreases phantom limb pain. Poster presentation at the 47th annual meeting of the Society for Neuroscience (SfN 2017), Washington, D.C., USA, Abstract # 499.04.

Petrini, F. P., Bumbasirevic, M., Valle, G., Ilic, V., Mijovic, P., Cvancara, P., Barberi, F., Bortolotti, D., Andreu, D., Divoux, J.-L., Lechler, K., Lesic, A., Mazic, S., Mijovic, B., Guiraud, D., Stieglitz, T., Asgeir, A., Micera, S., Raspopovic, S. (2019). Sensory feedback restoration in leg amputees improves walking speed, metabolic cost and phantom pain. Nature Medicine, 25: 1356–1363.

Poppendieck, W., Sossalla, A., Krob, M. O., Welsch, C., Nguyen, T. A., Gong, W., DiGiovanna, J., Micera, S., Merfeld, D. M. and Hoffmann, K. P. (2014). Development, manufacturing and application of double-sided flexible implantable microelectrodes. Biomed Microdevices 16(6): 837–850.

Raspopovic, S., Capogrosso, M., Petrini, F. M., Bonizzato, M., Rigosa, J., Pino, G. D., Carpaneto, J., Controzzi, M., Boretius, T., Fernandez, E., Granata, G., Oddo, C. M., Citi, L., Ciancio, A. L., Cipriani, C., Carrozza, M. C., Jensen, W., Guglielmelli, E., Stieglitz, T., Rossini, P. M. and Micera, S. (2014). Restoring Natural Sensory Feedback in Real-Time Bidirectional Hand Prostheses. Sci Transl Med 6(22): 222ra19, 10 pages.

Raspopovic, S., Petrini, F. M., Valle, G., Cvancara, P., Hiairrassaary, A., Guiraud, D., Alexandersson, A., Stieglitz, T., Micera, S. and Bumbarirevic, M. (2017). Bionic legs restore the dexterity, confidence and ownership in lower-limb amputees. Poster presentation at the 47th annual meeting of the Society for Neuroscience (SfN 2017), Washington, D.C., USA, Abstract # 718.06.

Rognini, G., Petrini, F. M., Raspopovic, S., Granata, G., Strauss, I., Valle, G., Solca, M., Bello-Ruiz, J., Herbelin, B., Mange, R., Di Iorio, R., Di Pino, G., Stieglitz, T., Rossini, P. M., Serino, A., Micera, S., Blanke, O. (2018). A multisensory bionic limb to achieve prosthesis embodiment and reduce distorted phantom limb perceptions. Neurol Neurosurg Psychiatry, jnnp-2018-318570.

Stieglitz, T. (2001). Flexible biomedical microdevices with double-sided electrode arrangements for neural applications, Sens Actuators A-Phys, A 90: 203–211.

Stieglitz, T., Boretius, T., Navarro, X., Badia, J., Guiraud, D., Divoux, J.-L., Mi-cera, S., Rossini, P. M., Yoshida, K., Harreby, K. R., Kundu, A. and Jensen, W. (2012). Development of a neurotechnological system for relieving phantom limb pain using transversal intrafascicular electrodes (TIME). Biomed Techn 57(6): 457–465.

Stieglitz, T. and Gross, M. (2002). Flexible BIOMEMS with electrode arrangements on front and back side as key component in neural prostheses and biohybrid systems, Sens Actuators B-Chem, B 83: 8–14.

Index

About the Editor

Winnie Jensen is currently working as a professor in neural engineering at the Department of Health science and Technology at Aalborg University in Denmark. She received her Ph.D. degree in bioengineering in 2001 from the same institution. Her research focus is on improving quality of life for people with impaired sensory or motor capabilities through development of innovative, implantable technological solutions that are inspired by human biology and by bridging animal and human research to create unique knowledge and novel rehabilitation systems.